The IMA Volumes in Mathematics and its Applications

Volume 133

Series Editors
Douglas N. Arnold   Fadil Santosa

Springer
New York
Berlin
Heidelberg
Hong Kong
London
Milan
Paris
Tokyo

Institute for Mathematics and its Applications (IMA)

The **Institute for Mathematics and its Applications** was established by a grant from the National Science Foundation to the University of Minnesota in 1982. The primary mission of the IMA is to foster research of a truly interdisciplinary nature, establishing links between mathematics of the highest caliber and important scientific and technological problems from other disciplines and industry. To this end, the IMA organizes a wide variety of programs, ranging from short intense workshops in areas of exceptional interest and opportunity to extensive thematic programs lasting a year. IMA Volumes are used to communicate results of these programs that we believe are of particular value to the broader scientific community.

The full list of IMA books can be found at the Web site of the Institute for Mathematics and its Applications:

http://www.ima.umn.edu/springer/full-list-volumes.html.

Douglas N. Arnold, Director of the IMA

* * * * * * * * * *

IMA ANNUAL PROGRAMS

1982–1983	Statistical and Continuum Approaches to Phase Transition
1983–1984	Mathematical Models for the Economics of Decentralized Resource Allocation
1984–1985	Continuum Physics and Partial Differential Equations
1985–1986	Stochastic Differential Equations and Their Applications
1986–1987	Scientific Computation
1987–1988	Applied Combinatorics
1988–1989	Nonlinear Waves
1989–1990	Dynamical Systems and Their Applications
1990–1991	Phase Transitions and Free Boundaries
1991–1992	Applied Linear Algebra
1992–1993	Control Theory and its Applications
1993–1994	Emerging Applications of Probability
1994–1995	Waves and Scattering
1995–1996	Mathematical Methods in Material Science
1996–1997	Mathematics of High Performance Computing
1997–1998	Emerging Applications of Dynamical Systems
1998–1999	Mathematics in Biology

Continued at the back

Peter J. Olver   Allen Tannenbaum
Editors

Mathematical Methods in Computer Vision

With 86 Figures

Springer

Peter J. Olver
School of Mathematics
University of Minnesota
127 Vincent Hall
206 Church St. S.E.
Minneapolis, MN 55455
USA
olver@ima.umn.edu

Allen Tannenbaum
Departments of Electrical and Computer and Biomedical Engineering
Georgia Institute of Technology
Atlanta, GA 30332-0250
USA
tannenba@ece.gatech.edu

Mathematics Subject Classification (2000): 68T50, 68U10, 65K10

Library of Congress Cataloging-in-Publication Data
Mathematical methods in computer vision / editors, Peter J. Olver, Allen Tannenbaum.
p. cm. — (The IMA volumes in mathematics and its applications ; 133)
Includes bibliographical references and index.

1. Computer vision—Mathematics. I. Olver, Peter J. II. Tannenbaum, Allen, 1953–
III. Series.
TA1634.M37 2003
006.3′7—dc21 2003042438

ISBN 978-1-4419-1826-0

Printed in the United States of America.

9 8 7 6 5 4 3 2 1

www.springer-ny.com

Springer-Verlag New York Berlin Heidelberg
A member of BertelsmannSpringer Science+Business Media GmbH

FOREWORD

This IMA Volume in Mathematics and its Applications

MATHEMATICAL METHODS IN COMPUTER VISION

contains papers presented at two successful one-week workshops: Image Processing and Low Level Vision and Image Analysis and High Level Vision. Both workshops were integral to the IMA annual program on Mathematics in Multimedia, 2000–2001. The first workshop, which took place October 16–20, 2000 was organized by Donald McClure (Brown University), Peter Olver (University of Minnesota), Pietro Perona (California Institute of Technology), and Allen Tannenbaum (Georgia Institute of Technology). Yali Amit (University of Chicago), Donald Geman (University of Massachusetts), Peter Olver (University of Minnesota), Allen Tannenbaum (Georgia Institute of Technology), Steven Zucker (Yale University) were organizers of the second workshop which was held November 13-17, 2000. We are grateful to the organizers for making the events successful.

We would like to thank Peter Olver and Allen Tannenbaum for their excellent work in editing the proceedings. We take this opportunity to thank the National Science Foundation and the Office of Naval Research for their support of the IMA.

Series Editors

Douglas N. Arnold, Director of the IMA

Fadil Santosa, Deputy Director of the IMA

PREFACE

This volume comprises some of the key work presented at two IMA Workshops on Computer Vision during fall of 2000. These workshops were devoted to exciting recent developments in the mathematical theory and practical applications in the field of computer vision and image processing. The workshop speakers included many of the leading researchers world wide, and drew a large and varied audience of mathematicians, engineers, computer scientists, mathematical biologists, psychologists, and physicists. The IMA proved to be very conducive to significant interactions, and initiation of new collaborative efforts as a result of these workshops as a part of the 2000–2001 IMA Multimedia program.

The first Workshop was devoted to "Image Processing and Low Level Vision" (September 2000). A primary goal of the Low-level Vision Workshop was to educate and interest mathematicians in the mathematical and scientific problems that arise in basic image processing. Recent years have seen significant advances in the application of sophisticated mathematical theories to the problems arising in image processing. As yet, even very low level visual processing by computers remains a challenging problem. Both planar and three-dimensional images are of importance, and these arise in a wide variety of applications, including medical imagery — ultrasound, nuclear magnetic resonance, X-ray computed topography, etc. — military and industrial imaging, and film restoration and animation. Basic issues include image smoothing and denoising, image enhancement, morphology, image compression, segmentation (determining boundaries of objects — including problems of camera distortion and partial occlusion).

Several mathematical approaches have emerged, including methods based on nonlinear partial differential equations, stochastic and statistical methods, and signal processing techniques, including wavelets and other transform theories. Partial differential equations are used to describe the evolution of shapes under curvature-controlled diffusions, providing a multiscale representation that is based upon curvature flows of fundamental importance in differential geometry. These methods have proven successful in noise reduction while maintaining edge retention. Applications to segmentation are based on a variational formulation of the method of "snakes" or active contours, in which an initial contour converges to the object boundary via a gradient descent flow based on a conformally Riemannian metric. Wavelets have applications to practical image compression methods, and texture characterization. Statistical methods such as the EM algorithm have been successfully applied to a variety of low level vision problems.

The theme of the second Workshop was "High-Level Vision." This took place at the IMA in November 2000. This workshop concentrated on mathematical and practical issues arising in the higher level processes in image analysis. These include object recognition, optical character and handwriting recognizers, printed-circuit board inspection systems, and quality control devices, motion detection, robotic control by visual feedback, theory of shape, reconstruction of objects from stereoscopic view and/or motion, and many others.

A number of the talks were devoted to shape theory, a topic of fundamental importance since it is the bottle-neck between high and low level vision. Thus "shape" formed the bridge between the two workshops on vision. The recent geometric partial differential equation methods have been essential in throwing new light on this very difficult problem area. These are based on certain conservation principles, Hamilton-Jacobi theory, and curvature driven flows which lead to a computational theory of shape which naturally characterizes the computational elements including protrusions, parts, bends, and seeds (which show where to place the components of a shape).

Stochastic processes, including Markov random fields, have been used in a Bayesian framework to incorporate prior constraints on a smoothness and the regularities of discontinuities into algorithms for image restoration and reconstruction. Such statistical methods also been used with great success in the field of speech recognition, which was the subject of a simultaneous program at the IMA during the fall semester. Interested readers may wish to consult the companion IMA volume containing the proceedings of the two speech workshops. Sequential decision theory has been used to develop algorithms for efficient identification of objects in a scene, including handwritten characters, roads in satellite imagery, and faces. Deformable templates have been used to automate the identification of structures, both normal and pathological, in medical imagery.

The papers listed below cover some of the key themes of the two workshops and were selected to highlight vision work with interesting mathematics as well as key applications. The paper "Analysis and Synthesis of Visual Images in the Brain" by Tai Sing Lee describes neurophysiological evidence which suggests that the early visual cortex participates in many levels of visual processing underlying the generation and the representation of the subjective visual world in our brain. The author argues that at each moment in time, one only perceives a very small fragment of the world through the retinas, yet the subjective perception of the visual world in front is rather clear, coherent and complete. One often sees things that are not even there because what one perceives is actually a 'virtual' visual world that is created in the mind. This virtual world is dynamic and plastic, and depends on the behavioral demands imposed on the subject and the statistics of our past experience.

The theme of the paper by Ian Dryden is "Statistical Shape Analysis in High-Level Vision." Here the author reviews some of the main aspects of statistical shape analysis and its use in high-level vision. He gives an introduction to shape and shape space, and then describes some common choices of shape metrics. Shape models are considered and their role as priors in high-level image analysis is described.

James Coughlan and Alan Yuille also consider statistical methods in their contribution "A Large Deviation Theory Analysis of Bayesian Tree Search." They argue that many perception, reasoning, and learning problems can be expressed in terms of Bayesian inference, requiring specification of a probability distribution on the ensemble of problem instances. The statistical model can be used for analyzing the expected complexity of algorithms and also the algorithm-independent limits of inference. The authors illustrate this by analyzing the problem of road detection.

Tryphon Georgiou, Peter Olver, and Allen Tannenbaum study the problem of "Maximal Entropy for Reconstruction of Back Projection Images" in their work. They point out that maximum entropy methods have proven to be a powerful tool for reconstructing data from incomplete measurements or in the presence of noise. The method is applied to the reconstruction of images from computed tomography data derived from back-projection over a finite set of angles.

Next, Ilya Pollack in his paper "Nonlinear Diffusions and Optimal Estimation" considers a nonlinear diffusion process known to be effective for image segmentation and binary classification problems. He shows how to optimally solve certain estimation problems, resulting in an efficient and exact method for solving the total variation minimization problem in one-dimension. This work is a natural outgrowth of some of the exciting developments in the field of nonlinear diffusion filtering of images and geometric scale-space methods that have taken place over the past decade.

The Mumford–Shah functional has become an example of the use of powerful mathematics to treat a key vision problem. Anthony Yezzi, Stefano Soatto, Andy Tsai, and Alan Willsky illustrate the versatility of this tool in their paper "The Mumford–Shah Functional: From Segmentation to Stereo." They first address the problem of simultaneous image segmentation and smoothing by approaching the Mumford–Shah paradigm from a curve evolution perspective. They let a set of deformable contours define the boundaries between regions in an image where the data is modeled by piecewise smooth functions, and employ a gradient flow to evolve these contours. Then the authors show how a related model may be used to segment multiple images of a 3D object by evolving a surface. Projections of this surface onto each 2D image gives rise to a family of contours that define boundaries between smooth regions within each image.

Steven Haker and Allen Tannenbaum consider the problems of image warping and registration using optimal transport theory in their paper "Monge–Kantorovich and Image Warping." Image registration is the process of establishing a common geometric reference frame between two or more data sets from the same or different imaging modalities possibly taken at different times. In the context of medical imaging and, in particular, image guided therapy, the registration problem consists of finding automated methods that align multiple data sets with each other and with the patient. In this paper, the authors propose a method of elastic registration based on the Monge–Kantorovich problem of optimal mass transport.

Finally, Ernst Dickmanns presents some of his pioneering work in active vision in his paper "Expectation-based, Multi-focal, Saccadic Vision (Understanding Dynamic Scenes Observed From a Moving Platform)." Professor Dickmanns presents his third-generation dynamic vision system based on his experience with autonomous road vehicles. He describes the "MarVEye" camera configuration which is realized with three to four conventional miniature CCD-TV-cameras mounted on a high bandwidth pan-and-tilt platform allows new levels of performance. A coherent form of representation of dynamic scenes containing multiple independently moving objects has been achieved through homogeneous coordinate transformations as edges in a scene tree. The author presents his overall system based on COTS-PC-hard- and software, visual/inertial data fusion and object-oriented programming.

We hope that the selection of papers in this volume will give the reader a taste of the broad range of exciting mathematical and practical developments in computer vision and related areas of contemporary research.

We would like to thank our fellow organizers Yali Amit, Donald Geman, Stuart Geman, Don McLure, Pietro Perona, and Steven Zucker for all of their efforts in making these workshops a success. Special thanks goes to the IMA Director, Willard Miller, for all of his help as well as to the professional staff of the IMA.

Peter J. Olver
School of Mathematics
University of Minnesota

Allen Tannenbaum
Departments of Electrical and Computer and Biomedical Engineering
Georgia Institute of Technology

CONTENTS

A LARGE DEVIATION THEORY ANALYSIS OF BAYESIAN TREE SEARCH

JAMES M. COUGHLAN* AND ALAN L. YUILLE*

Abstract. Many perception, reasoning, and learning problems can be expressed as Bayesian inference. We point out that formulating a problem as Bayesian inference implies specifying a probability distribution on the ensemble of problem instances. This ensemble can be used for analyzing the expected complexity of algorithms and also the algorithm-independent limits of inference. We illustrate this by analyzing the problem of road detection, formulated as tree search, by Geman and Jedynak [6]. This analysis uses large deviation theory to put bounds on the probability of rare events, such as exploring wrong branches of the tree in depth. We prove that the expected convergence is linear in the size of the road (i.e. depth of the tree) even though the worst case performance is exponential.

Key words. Bayesian inference, large deviation theory, tree search.

1. Introduction. Many problems in vision, speech, reasoning, and other sensory and control modalities can be formulated as Bayesian inference [9]. It is important to understand the complexities of algorithms which can perform these inferences.

We point out that formulating a problem as Bayesian inference implies specifying a probability distribution on the *ensemble of problem instances.* More formally, in Bayesian inference the goal is to estimate a quantity x from data y by using the *posterior* distribution $P(x|y)$. Constructing this posterior requires specifying a *likelihood function* $P(y|x)$ and a *prior* distribution $P(x)$. From these distributions we can construct a distribution $P(x, y)$ on the ensemble of problem instances (x, y). See Figure 4 for samples from a particular ensemble for road tracking.

There are two main advantages to analyzing the performance of an algorithm over the ensemble of problem instances. Firstly, it allows us to determine the behavior of the algorithm for typical problem instances (i.e. those which occur with non-negligible probability) and means that we may not have to deal with worst case situations (because they may have arbitrarily small probabilities). Secondly, having a distribution over the ensemble of problem instances also enables us to quantify the accuracy of the estimates found by the algorithm.

In this paper, we illustrate these advantages by analyzing the complexity of an algorithm proposed by Geman and Jedynak [6] for detecting roads in aerial images. Geman and Jedynak formulated the problem as Bayesian maximum a posteriori (MAP) estimation. This reduces the problem to tree search. In this paper we analyze the complexity of a variant of the Geman and Jedynak algorithm (this variant was proposed by the authors in [18]).

*Smith-Kettlewell Eye Research Institute, 2318 Fillmore St., San Francisco, CA 94115.

The complexity, and performance, of the algorithm depends on the probability distributions which characterize the ensemble of problem instances (i.e. the likelihood function and the prior). For a large class of ensembles, we are able to prove that the expected complexity is linear in the length N of the road (by contrast, the worst case complexity is exponential in N). We are also able to put bounds on the expected errors made by the algorithm.

We emphasize that we are concerned with the ability of the algorithm to detect the road path *which may not necessarily correspond to the MAP estimate.* For any problem instance there are three important paths: (i) the true road path, (ii) the MAP estimate of the true road path, and (iii) the path found by the algorithm. In this paper we are concerned with the difference between (iii) and (i) only.

These results complement our previous work [20, 21] which used this ensemble concept to analyze the *algorithm-independent* performance of MAP estimation on problems of this type (i.e. we evaluated errors between the MAP estimate (ii) and the true road path (i)). In particular, we derived an *order parameter* K_B which is a function of the ensemble. We proved that if $K_B < 0$ then it is impossible to detect the true road by *any algorithm.* (Not surprisingly, our results in this paper on linear expected complexity only apply to ensembles for which $K_B > 0$).

Another advantage of the ensemble concept is illustrated by our choice of a *heuristic* A^* algorithm [18]. A* algorithms [11, 16, 13] search trees and/or graphs using a *heuristic* to estimate future rewards. If we are using a Bayesian ensemble then the probability distributions can be used to generate heuristics.

Technically, our proofs make use of *large deviation theory* [7]. In particular we use Sanov's theorem, see [4], to put bounds on the probability of rare events. We note that many results on statistical learning theory [15] are derived using similar techniques from large deviation theory.

Finally, we would like to mention related work on optimization which uses the concept of ensembles.

Firstly, Karp and Pearl [8, 11] provided a theoretical analysis of convergence rates of A* search by considering an ensemble of problem instances. They studied a binary tree where the rewards for each arc were 0 or 1 and were specified by a probability p. They then obtained the complexity of algorithms for finding the minimum cost path. This work has some similarities to ours but their formulation is not Bayesian, their heuristics for A* algorithms are different, and large deviation theory is not used. Their work was an inspiration for us and we provided an analysis of a block pruning algorithm motivated by them in [17].

Secondly, there are some recent studies showing that order parameters exist for NP-complete problems and that these problems can be easy to solve for certain values of the order parameters [1, 14]. This work involves analyzing ensembles of problem instances. But the distribution of instances

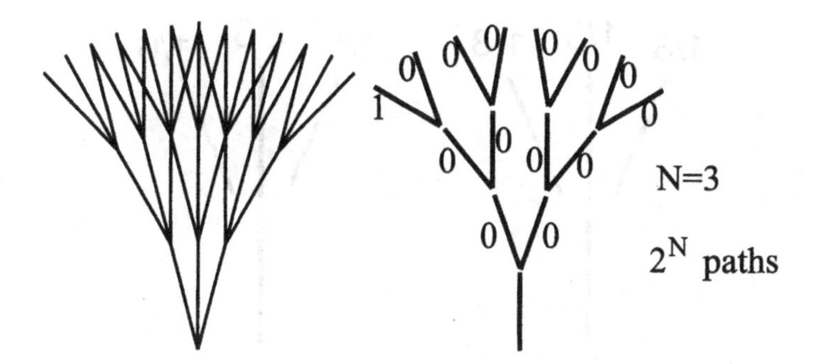

FIG. 1. *Left Panel: Geman and Jedynak's tree structure with a branching factor of* $Q = 3$. *Right Panel: The worst case complexity is exponential because, for some problem instances, the best path is determined only by the reward of the final arc segment. In this panel* $Q = 2$ *and all the* 2^N *paths have to be examined.*

in the ensemble is typically assumed to be uniform and is not derived from Bayesian methods.

The structure of this paper is as follows. In Section (2) we formulate the road tracking problem and introduce A*. Section (3) gives complexity results for a special choice of A* heuristic (making use of Sanov's theorem). These results can be extended to other heuristics [2]. Section (4) proves that sorting the queue for A* takes constant expected time per sort operation.

2. Problem formulation.

2.1. Tree search: the Geman and Jedynak model. Many problems in artificial intelligence can be formulated as tree search ([16, 11, 13]). We now study a specific example of this class of problem.

Geman and Jedynak [6] formulate road detection as tree search in a Q-nary tree, see Figure 1. The starting point and initial direction are specified and there are Q^N possible distinct paths down the tree. The goal is to find the road path (whose statistical properties differ from those of the non-road paths). The worst case complexity for this problem is exponential but, as we will prove, in many circumstances it is possible to detect a good approximation to the road path in linear expected time. Our analysis involves considering an ensemble of problem instances.

More formally, a road hypothesis, or path, consists of a set of connected straight-line *segments*. We can represent a path by a sequence of moves $\{t_i\}$ on the tree. Each move t_i belongs to an *alphabet* $\{b_\nu\}$ of size Q. For example, the simplest case studied by Geman and Jedynak sets $Q = 3$ with an alphabet b_1, b_2, b_3 corresponding to the decisions: (i) b_1 – go straight (0 degrees), (ii) b_2 – go left (-5 degrees), or (iii) b_3 – go right (+ 5 degrees).

Each tree will contain a *target path* which corresponds to the road to be detected. This path is sampled from a prior probability distribution $P(\{t_i\}) = \prod_{i=1}^{N} P_{\Delta G}(t_i)$, where $P_{\Delta G}(.)$ is the geometric transition proba-

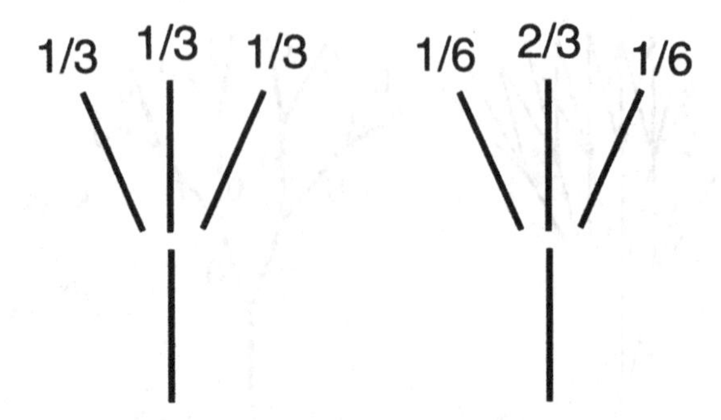

FIG. 2. *Different priors for the geometry. (Left Panel) the probabilities of turning left, right, or straight are* 1/3. *(Right Panel) the probability of going straight is* 2/3 *and the probabilities of turning right or left are* 1/6 *each, biasing towards straighter paths.*

bility. For our $Q = 3$ example, we may choose to go straight, left or right with equal probability (i.e. $P_{\Delta G}(b_1) = P_{\Delta G}(b_2) = P_{\Delta G}(b_3) = 1/3$), see Figure 2.

A sequence of moves $\{t_i : i = 1, ..., N\}$ determines a path of segments $X = (x_1, ..., x_N)$ (ie. these segments form a connected path from the top of the tree to the bottom). Conversely, a consistent path X determines a sequence of moves. (This also applies to subpaths). Let χ denote all the segments of the tree. So a path X is a connected subset of χ. The set of segments not on the path is the complement $\chi \setminus X$.

There is an *observation* y_x for each segment $x \in \chi$ of the tree. The set of all observations is $Y = \{y_x : x \in \chi\}$. The values of the observations belong to an alphabet $\{a_\mu\}$ of size J. An observation y_x is drawn from a distribution $P_{on}(.)$ if the segment is on the target path (ie. if $x \in X$). If not, it is drawn from $P_{off}(.)$. For any path $\{t_i\}$ through the tree, with segments $\{x_i\}$, we have a corresponding set of observations $\{y_{x_i}\}$. (The observation y_x is the response to a non-linear filter, evaluated on the image at segment x, which is designed to detect straight road segments. The filter is trained on examples of on-road and off-road segments to determine empirical distributions $P_{on}(y)$ and $P_{off}(y)$, as described in [6]. See Figure 3 for examples of distributions P_{on}, P_{off}, taken from [10]).

This determines the likelihood function $P(Y|X)$:

$$P(Y|X) = \prod_{x \in X} P_{on}(y_x) \prod_{x \in \chi \setminus X} P_{off}(y_x), \tag{2.1}$$

which we can re-express as:

$$P(Y|X) = \prod_{i=1,\dots,N} \frac{P_{on}(y_{x_i})}{P_{off}(y_{x_i})} \prod_{x \in \chi} P_{off}(y_x) = \prod_{i=1,\dots,N} \frac{P_{on}(y_{x_i})}{P_{off}(y_{x_i})} F(Y). \tag{2.2}$$

where $F(Y) = \prod_{x \in \chi} P_{off}(y_x)$ is independent of the target path X.

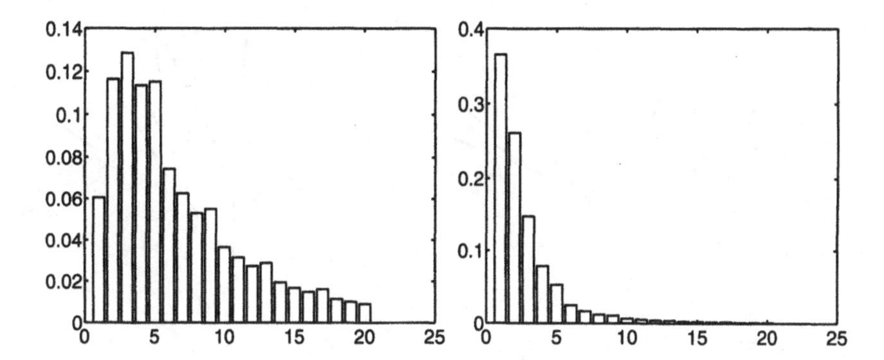

FIG. 3. *The quantized distributions $P_{on}(y)$ (Left) and $P_{off}(y)$ (Right), where $y = \left|\vec{\nabla} I(\mathbf{x})\right|$, learned from image data. Observe that, not surprisingly, $\left|\vec{\nabla} I(\mathbf{x})\right|$ is likely to take larger values on an edge rather than off an edge.*

We formulate the problem as MAP estimation to find the mode of the posterior distribution $P(X|Y) = P(Y|X)P(X)/P(Y)$ where $P(X) = \prod_{i=1}^{N} P_{\Delta G}(t_i)$ (where $\{t_i\}$ is the sequence of moves that generates the path X).

Then MAP estimation for X is equivalent to maximizing

$$P(Y|X)P(X) = \prod_{i=1}^{N} P_{\Delta G}(t_i) \prod_{i=1,\dots,N} \frac{P_{on}(y_{x_i})}{P_{off}(y_{x_i})} F(Y). \tag{2.3}$$

This is equivalent to finding the path $\{t_i\}$ with observations $\{y_i\}$ which maximizes the following *reward function*:

$$r(\{t_i\},\{y_i\}) = \sum_{i=1}^{N} \log\left\{\frac{P_{on}(y_i)}{P_{off}(y_i)}\right\} + \sum_{i=1}^{N} \log\left\{\frac{P_{\Delta G}(t_i)}{U(t_i)}\right\}, \tag{2.4}$$

where y_i is shorthand for y_{x_i}, $U(.)$ is the uniform distribution (i.e. $U(b_\nu) = 1/Q\ \forall \nu$) and so $\sum_{i=1}^{N} \log U(t_i) = -N \log Q$ which is a constant. The introduction of $U(.)$ helps simplify the analysis in the following subsections.

Observe that the reward of a particular path depends only on the variables $\{t_i\}, \{y_i\}$ which define the path (the moves and the observations). This is because the factor $F(Y)$ in $P(Y|X)$ is independent of X and can be ignored (i.e. it does not affect which path is most probable).

For any path (or subpath) in the tree of length n we can re-express the reward function $r(\{t_i\}, \{y_i\})$ as:

$$r(\{t_i\},\{y_i\}) = n\vec{\phi}\cdot\vec{\alpha} + n\vec{\psi}\cdot\vec{\beta}, \tag{2.5}$$

where $\vec{\alpha}$ and $\vec{\beta}$ have components:

$$\alpha_\mu = \log\frac{P_{on}(a_\mu)}{P_{off}(a_\mu)},\ \mu = 1,\dots,J, \quad \beta_\nu = \log\frac{P_{\Delta G}(b_\nu)}{U(b_\nu)},\ \nu = 1,\dots,Q. \tag{2.6}$$

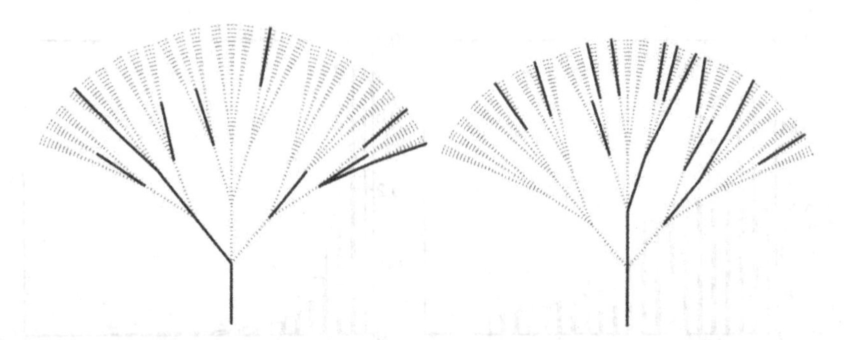

FIG. 4. *Samples from the Bayesian ensemble. Simulated road tracking problem where dark lines indicate strong edge responses and dashed lines specify weak responses. The data was generated by stochastic sampling using a simplified version of the models analyzed in this paper. In both examples there is only one strong candidate for the best path (the continuous dark line) but chance fluctuations have created subpaths in the noise with strong edge responses.*

and $\vec{\phi}$ and $\vec{\psi}$ are normalized histograms, or *types* [4], with components (with $\delta_{i,j}$ denoting the Kronecker delta function):

$$(2.7) \quad \phi_\mu = \frac{1}{n}\sum_{i=1}^{n} \delta_{y_i,a_\mu},\ \mu = 1, ..., J\ ,\ \psi_\nu = \frac{1}{n}\sum_{i=1}^{n} \delta_{t_i,b_\nu}\ \nu = 1, ..., Q.$$

We now illustrate the Bayesian ensemble by Figure 4 which consists of two samples from the ensemble. In these cases the target path is easily detectable (i.e. the target reward is higher than the reward of any other path) but noise fluctuations mean that some subpaths may distract the algorithm from the target. In other ensembles, the target path may be far harder to detect.

2.2. Can the task be solved? Distractor paths. The goal is to detect the target path by selecting the path with highest reward. But the MAP estimate may not necessarily correspond to the target path. In this subsection, we specify conditions which ensure that the MAP estimate is expected to be significantly *similar* to the target path. Unless these conditions are satisfied it will be *impossible to find the target path by any algorithm.*

The tree contains one target path and $Q^N - 1$ distractor paths. We categorize the distractor paths by the stage at which they diverge from the target path, see Figure 5. For example, at the first branch point the target path lies on only one of the Q branches and there are $Q-1$ false branches which generate the first set of false paths F_1. Now consider all the $Q-1$ false branches at the second target branch, these generate set F_2. As we follow along the true path we generate sets F_i of size $(Q-1)Q^{N-i}$. The set of all paths is therefore the target path plus the union of the F_i $(i = 1, \ldots, N)$.

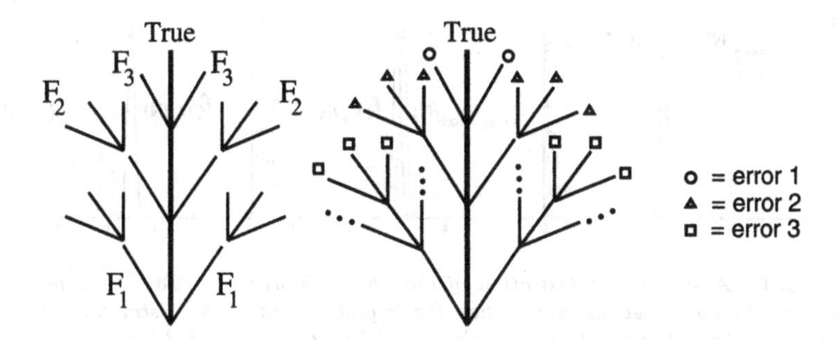

FIG. 5. *Left: Given a specified target path (the straight line shown in bold in this case) we can divide the set of paths up into N subsets $F_1, ..., F_N$ as shown here. Paths in F_1 have no overlap with the target path. Paths in F_2 overlap with one segment, and so on. Intuitively, we can think of this as an onion where we peel off paths stage by stage. Right: When paths leave the target path they make errors which we characterize by the number of false segments. For example, a path in F_1 has error N, a path in F_i has error $N+1-i$.*

To determine whether the target path can be found by MAP estimation we must consider the probability that one of the distractor paths has higher reward than the target path. For example, consider the probability distribution $\hat{P}_{1,max}(r_{max}/N)$ of the *maximum* reward (normalized by N) of all the paths in F_1. We can compare this to the probability distribution of the (normalized) reward $\hat{P}_T(r_T/N)$ of the target path. In related work [21], we use techniques similar to Sanov's theorem to estimate these quantities and to show that there is a phase transition depending on a parameter K given by:

$$K = D(P_{on}||P_{off}) + D(P_{\Delta G}||U) - \log Q, \tag{2.8}$$

where $D(P_{on}|P_{off}) = \sum_y P_{on}(y) \log \frac{P_{on}(y)}{P_{off}(y)}$ is the Kullback-Leibler divergence between P_{on} and P_{off}.

If $K > 0$ then the probability distribution for the target path reward $\hat{P}_T(r_T/N)$ lies to the right of the distribution $\hat{P}_{1,max}(r_{max}/N)$ of the maximum reward of paths in F_1, see Figure 6 left panel, and it is straightforward to detect the target path. At $K \approx 0$ the two distributions overlap and it becomes hard to detect the target path, see Figure 6 center panel. But if $K < 0$, then $\hat{P}_{1,max}(r_{max}/N)$ is to the right of $\hat{P}_T(r_T/N)$, see Figure 6, and it is impossible to detect the target path.

To get intuition for K we consider its three terms. The first term, $D(P_{on}||P_{off})$, is a measure of how effective the local filter cues are for detecting the target. If $P_{on} = P_{off}$ then $D(P_{on}||P_{off}) = 0$ and the local cues are useless. The second term $D(P_{\Delta G}||U)$ is a measure of how much prior knowledge we have about the probable shape of the target (setting $P_{\Delta G} = U$ means we have no prior information). Finally, $\log Q$ is a measure of how many distractor paths there are. Therefore K becomes larger (and

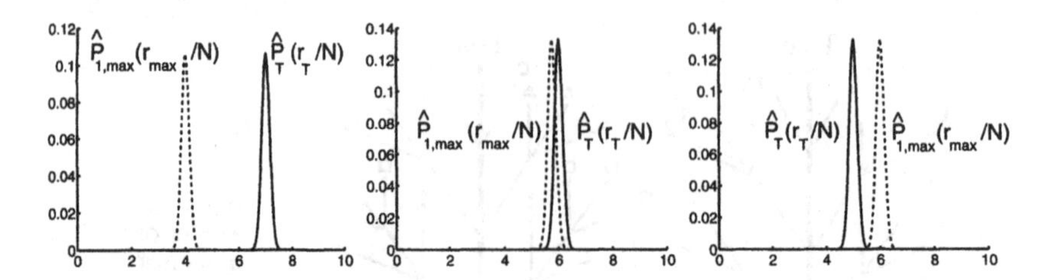

FIG. 6. *A schematic illustration of the Phase Transition. (Top left panel) The reward of the target path is higher than the largest reward of the distractor paths (so detection is straightforward). (Top right panel) The task becomes difficult because the target reward and the best distractor reward are very similar. (Bottom panel) Detecting the target path becomes impossible because its reward is lower than the best distractor path. The horizontal axis labels the normalized reward.*

so target detection becomes easier) with better filter detectors, more prior knowledge, and fewer distractor paths.

In related work [20] we used Sanov's theorem to show that the expected number of paths in F_1 with rewards greater than the target path behaves as 2^{-NK_B}, where the *order parameter* K_B is defined by:

$$K_B = 2B(P_{on}, P_{off}) + 2B(P_{\Delta G}, U) - \log Q, \tag{2.9}$$

where $B(P, Q) = -\log \sum_{i=1}^{m} (p_i)^{1/2}(q_i)^{1/2}$ is the Bhattacharyya bound between the distributions $P = \{p_i\}$ and $Q = \{q_i\}$. Once again, there is a change in behavior as K_B changes sign. For $K_B > 0$, we expect there to be no paths in F_1 with rewards greater than the target path.

Our analysis of the A* algorithm will proceed in the regime where $K > 0$ and $K_B > 0$. There is little purpose in estimating how fast one can compute the MAP estimator unless one is sure that the estimator is detecting a good approximation to the correct target. Our results, see section (3), will require an additional condition to hold, see Theorems 6 and 7, which will ensure that $K_B > 0$. There will, however, be situations where the target is detectable (ie. $K_B > 0$) but where we cannot prove expected linear convergence. More specifically, we can express $K_B = \{\psi_1 - \log Q\} + \psi_2$ where ψ_1, ψ_2 are positive quantities which are defined in Theorem 3. Our complexity proofs apply provided $\psi_1 > \log Q$. If $\psi_1 < \log Q$ but $\psi_1 + \psi_2 > \log Q$ then the target path is detectable but we can say nothing about the complexity of the algorithms.

2.3. A*. The A* graph search algorithm [11, 16, 13] is used to find a path of maximum reward between a start node A and a goal node B in a graph, see Figure 7. The reward of a particular path is the sum of the rewards of each edge traversed. The A* procedure maintains a tree of partial paths already explored, and computes a measure f of the "promise" of each partial path (i.e. leaf in the search tree). New paths are considered by extending the most promising node one step. The measure

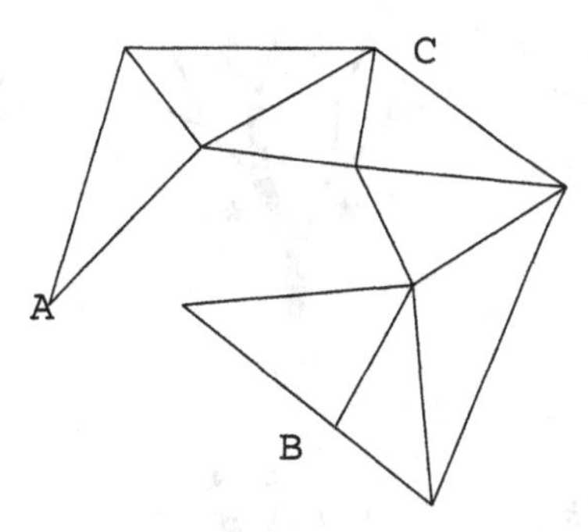

FIG. 7. *The A* algorithm tries to find the path from A to B with highest reward. For a partial path AC the algorithm stores g(C), the best reward to go from A to C, and an overestimate h(C) of the reward to go from C to B.*

f for any node C is defined as $f(C) = g(C) + h(C)$, where $g(C)$ is the best cumulative reward found so far from A to C and $h(C)$ is an overestimate of the remaining reward from C to B. The closer this overestimate is to the true reward then the faster the algorithm will run. We will refer to the value of f as the *A^* reward* in contrast with the reward function of equation (2.5).

It is straightforward to prove that A* is guaranteed to converge to the correct result provided the heuristic $h(.)$ is an upper bound for the true reward from all nodes C to the goal node B. A heuristic satisfying these conditions is called *admissible.* Conversely, a heuristic which does not satisfy them is called *inadmissible.* The word "inadmissible" is a technical term only and *does not* imply that inadmissible heuristics are inferior to admissible ones. In fact, as we show in this paper, algorithms using inadmissible heuristics can converge rapidly to good approximations to the correct result. Conversely, as discussed below, algorithms with admissible heuristics may be slow to converge.

In this paper, we consider inadmissible heuristics. We set the heuristic reward to be $H_L + H_P$ for each unexplored segment (where $_L$ and $_P$ label the likelihood and the geometric prior respectively). Thus a subpath starting at the origin of length M will have heuristic reward of $(N - M)(H_L + H_P)$. We will drop the $N(H_L + H_P)$ term, which is the same for all paths, and simply use $-M(H_L + H_P)$ as the heuristic.

We consider the A* rewards of two partial paths, one of length m segments that overlaps completely with the target path, and the other of length n that does not overlap at all with the target path. The A* rewards of these paths are denoted by $S_{on}(m)$ and $S_{off}(n)$ and are given by:

$$\begin{aligned} S_{on}(m) &= m\{\vec{\phi}^{on} \cdot \vec{\alpha} - H_L\} + m\{\vec{\psi}^{on} \cdot \vec{\beta} - H_P\}, \\ S_{off}(n) &= n\{\vec{\phi}^{off} \cdot \vec{\alpha} - H_L\} + n\{\vec{\psi}^{off} \cdot \vec{\beta} - H_P\}, \end{aligned} \tag{2.10}$$

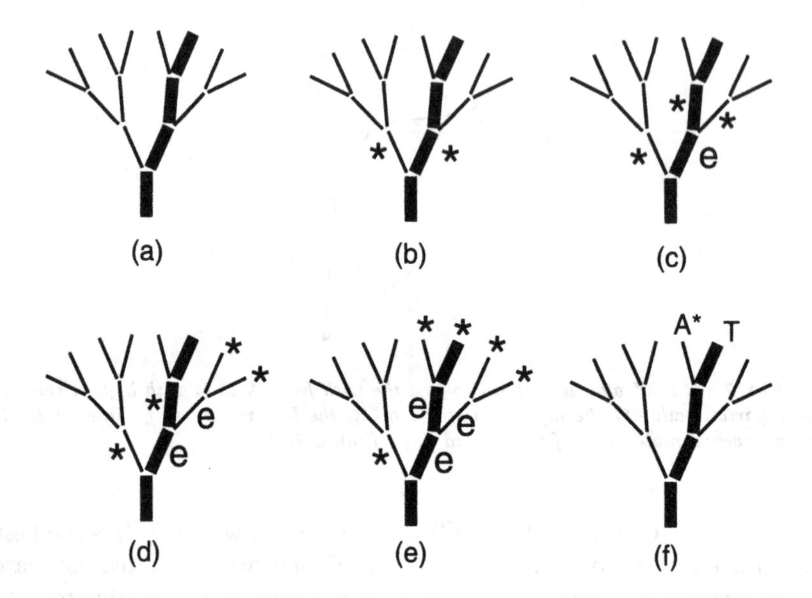

FIG. 8. *An A* search sequence. Panel (a) shows the target path in bold. Panel (b) shows the two segments added to the queue (marked by asterisks) at the start of the search. Panel (c) explores the right segment, removes it from the queue (label e), and adds its two children to the queue. The search continues in panels (d,e) where explored segments are eliminated from the queue (and labelled e) and their children are added to the queue (labelled with asterisks). Panel (f) A* converges to a solution (A*) that is one segment away from the target path (T).*

where $\phi_\mu^{on} = \frac{1}{m}\sum_i \delta_{y_i,a_\mu}$ and $\psi_\nu^{on} = \frac{1}{m}\sum_i \delta_{t_i,b_\nu}$ for the on path segments (and similarly for the off path segments).

The effect of this heuristic is to encourage us to explore subpaths provided their reward per segment is greater than $H_L + H_P$. Note that the larger the value of $H_L + H_P$ the more the algorithm will favor a breadth-first strategy [16] of exploring the tree (because few long subpaths will have rewards per segment exceeding $H_L + H_P$). Breadth-first search is a conservative strategy which will find the best solution but may take a long time to do so. In general, the smaller the heuristic then the faster the search time but the greater the possibility of error. In particular, if $H_L + H_P > \max_{\mu\in\{1,..,J\}} \alpha_\mu + \max_{\nu\in\{1,...,Q\}} \beta_\nu$ then the heuristic is admissible and the search is guaranteed to converge to the path with highest reward [11, 16, 13]. (But an algorithm which uses this heuristic is likely to be slow.) We will be considering inadmissible heuristics (i.e. $H_L + H_P < \max_{\mu\in\{1,..,J\}} \alpha_\mu + \max_{\nu\in\{1,...,Q\}} \beta_\nu$) which we expect to be quicker than admissible heuristics but which will have more errors (our theorems quantify these statements).

There is a close connection between heuristics for A* search and the general issue of pruning search algorithms. In previous work [19] we analyzed the results of a search algorithm which pruned out paths whose rewards fell below a critical value (corresponding to a heuristic). In related work, we explored the use of pruning heuristics for speeding up dynamic programming algorithms for detecting hand outlines in real images [3].

We will first prove convergence results for a specific choice of H_L, H_P and later generalize to a larger set of values.

3. Convergence proof with Bhattacharyya Heuristics. This section will prove convergence results for a specific choice of heuristic which we call the *Bhattacharyya Heuristic.* The next section will generalize the results to a larger class (for which the proofs are more complicated). First we need to say something about the choice of heuristics.

We need to choose a heuristic so that it is smaller than the reward (per unit length) that we would get from the target path and larger than the reward for a distractor path. This means, see equation (2.10), that the A* rewards will tend to be positive for the target and negative for the distractor paths (so the algorithm will prefer to explore the target). Let us consider the reward H_L only (the analysis is similar for H_P). The expected reward for the target path is $D(P_{on}||P_{off}) = \sum_y P_{on}(y) \log \frac{P_{on}(y)}{P_{off}(y)}$ and for the distractor path it is $-D(P_{off}||P_{on}) = \sum_y P_{off}(y) \log \frac{P_{on}(y)}{P_{off}(y)}$. Therefore we want to select H_L so that $-D(P_{off}||P_{on}) < H_L < D(P_{on}||P_{off})$.

Our complexity results will be obtained using Sanov's theorem [4] to estimate the probability that the algorithms wastes time searching distractor paths. In order to use Sanov's theorem, it is convenient to think of the heuristic as the expected reward of data distributed according to the mixture of P_{on}, P_{off} given by $P_\lambda(y) = P_{on}^{1-\lambda}(y) P_{off}^{\lambda}(y)/Z[\lambda]$ (where $Z[\lambda]$ is a normalization constant). In this section we will consider the special case where $\lambda = 1/2$. This gives the *Bhattacharyya heuristic* $H_L^* = \sum_y P_{\lambda=1/2}(y) \log \frac{P_{on}(y)}{P_{off}(y)}$. (We give it this name because the distribution $P_{\lambda=1/2}$ is associated with the Bhattacharyya bound in statistics [12]). By setting $\lambda = 1/2$ we are essentially choosing a heuristic midway between the target and distractors (generated by P_{on} and P_{off} respectively). (In [2] we will extend the results to deal with other values of λ.)

The Bhattacharyya heuristic is special in two ways. Firstly, it simplifies the analysis. Secondly, and more importantly, we can prove stronger results about convergence if the Bhattacharyya heuristic is used (although this may reflect limitations in our proofs). As we show in [2], if the algorithm converges using one of the alternative heuristics then it will also converges with the Bhattacharyya heuristic, but the reverse is not necessarily true.

The Bhattacharyya heuristics H_L^*, H_P^* are the expected rewards per segment:

$$H_L^* = \vec{\phi}_{Bh} \cdot \vec{\alpha}, \qquad H_P^* = \vec{\psi}_{Bh} \cdot \vec{\beta}, \tag{3.1}$$

with respect to the distributions ϕ_{Bh}, ψ_{Bh}:

$$\phi_{Bh}(y) = \frac{\{P_{on}(y)\}^{1/2}\{P_{off}(y)\}^{1/2}}{Z_\phi}, \qquad (3.2)$$
$$\psi_{Bh}(t) = \frac{\{P_{\Delta G}(t)\}^{1/2}\{U(t)\}^{1/2}}{Z_\psi},$$

where Z_ϕ, Z_ψ are normalization constants.

We first put an upper bound on the probability that any completely false segment is searched.

Let $A_{n,i}$ be the set of subpaths of length n that belong to F_i. Then we have the following result:

THEOREM 1. *The probability that A^* searches the last segment of a particular subpath in $A_{n,i}$ is less than or equal to $Pr\{\exists\ m : S_{off}(n) \geq S_{on}(m)\}$.*

Proof. By definition of A*, a **necessary** condition for the segment to be searched is that its A* reward (including the heuristic) is better than the A* reward of at least one segment on the target path. This is because the A* algorithm always maintains a queue of nodes to explore and searches the node segment with highest reward. The algorithm is initialized at the start of the target path and so an element of the target path will always lie in the queue of nodes that A* considers searching. (This condition is not sufficient to ensure that the segment is searched – so we are only obtaining an upper bound). □

We now bound $Pr\{\exists\ m : S_{off}(n) \geq S_{on}(m)\}$ by something we can evaluate.

THEOREM 2.
$$Pr\{\exists\ m : S_{off}(n) \geq S_{on}(m)\} \leq \sum_{m=0}^{\infty} Pr\{S_{off}(n) \geq S_{on}(m)\}.$$

Proof. Boole's inequality. □

We now proceed to find a bound on $Pr\{S_{off}(n) \geq S_{on}(m)\}$. This is done using Sanov's theorem (see [2] for details). It will show that this probability falls-off exponentially with n, m (provided certain parameters are positive).

THEOREM 3.
$$Pr\{S_{off}(n) \geq S_{on}(m)\} \leq \{(n+1)(m+1)\}^{J^2Q^2} 2^{-(n\Psi_1 + m\Psi_2)},$$
where $\Psi_1 = D(\vec{\phi}_{Bh}||P_{off}) + D(\vec{\psi}_{Bh}||U)$ *and* $\Psi_2 = D(\vec{\phi}_{Bh}||P_{on}) + D(\vec{\psi}_{Bh}||P_{\Delta G})$.

Proof. The proof is an application of Sanov's theorem, see [2], applied to the product space of types of $P_{on}, P_{off}, P_{\Delta G}, U$. Define:

$$E = \{(\vec{\phi}^{off}, \vec{\psi}^{off}, \vec{\phi}^{on}, \vec{\psi}^{on}) : n\{\vec{\phi}^{off} \cdot \vec{\alpha} - H_L^* + \vec{\psi}^{off} \cdot \vec{\beta} - H_p^*\} \qquad (3.3)$$
$$\geq m\{\vec{\phi}^{on} \cdot \vec{\alpha} - H_L^* + \vec{\psi}^{on} \cdot \vec{\beta} - H_p^*\}\}.$$

(i.e. E is the set of all histograms corresponding to partial off paths with higher A* reward than the partial on path).

Sanov's theorem gives a bound in terms of the $\phi^{\text{off}}, \psi^{off}, \phi^{on}, \psi^{on}$ that minimize:

$$
\begin{aligned}
f(\vec{\phi}^{off}, \vec{\psi}^{off}, \vec{\phi}^{on}, \vec{\psi}^{on}) = {} & nD(\vec{\phi}^{off}||P_{off}) + nD(\vec{\psi}^{off}||U) \\
& + mD(\vec{\phi}^{on}||P_{on}) + mD(\vec{\psi}^{on}||P_{\Delta G}) \\
& + \tau_1\Big\{\sum_y \phi^{off}(y) - 1\Big\} + \tau_2\Big\{\sum_t \psi^{off}(t) - 1\Big\} \\
& + \tau_3\Big\{\sum_y \phi^{on}(y) - 1\Big\} + \tau_4\Big\{\sum_t \psi^{on}(t) - 1\Big\} \\
& + \gamma\{m\{\vec{\phi}^{on}\cdot\vec{\alpha} - H_L^* + \vec{\psi}^{on}\cdot\vec{\beta} - H_p^*\} \\
& - n\{\vec{\phi}^{off}\cdot\vec{\alpha} - H_L^* + \vec{\psi}^{off}\cdot\vec{\beta} - H_p^*\}\},
\end{aligned} \tag{3.4}
$$

where the τ's and γ are Lagrange multipliers. This function $f(.,.,.,.)$ is known to be convex so there is a unique minimum. Observe that $f(....)$ consists of four terms of form $nD(\vec{\phi}^{off}||P_{off}) + \tau_1\{\sum_y \phi^{off}(y) - 1\} - n\gamma\vec{\phi}^{off}\cdot\vec{\alpha}$ which are coupled only by shared constants. These terms can be minimized separately to give:

$$
\begin{aligned}
\vec{\phi}^{off*} &= \frac{P_{on}^{\gamma} P_{off}^{1-\gamma}}{Z[1-\gamma]}, & \vec{\phi}^{on*} &= \frac{P_{on}^{1-\gamma} P_{off}^{\gamma}}{Z[\gamma]}, \\
\vec{\psi}^{off*} &= \frac{P_{\Delta G}^{\gamma} U^{1-\gamma}}{Z_2[1-\gamma]}, & \vec{\psi}^{on*} &= \frac{P_{\Delta G}^{1-\gamma} U^{\gamma}}{Z_2[\gamma]},
\end{aligned} \tag{3.5}
$$

subject to the constraint given by equation (3.3).

By inspection, the unique solution occurs when $\gamma = 1/2$. In this case:

$$
\vec{\phi}^{off*}\cdot\vec{\alpha} = H_L^* = \vec{\phi}^{on*}\cdot\vec{\alpha}, \qquad \vec{\psi}^{off*}\cdot\vec{\beta} = H_P^* = \vec{\psi}^{on*}\cdot\vec{\beta}. \tag{3.6}
$$

The solution occurs at $\vec{\phi}^{on*} = \vec{\phi}^{off*} = \vec{\phi}_{Bh}$ and at $\vec{\psi}^{on*} = \vec{\psi}^{off*} = \vec{\psi}_{Bh}$. Substituting into the Sanov bound gives the result. □

From Theorem 3, it is a direct summation, and application of Theorem 2, to obtain:

THEOREM 4. *$Pr\{\exists\, m : S_{off}(n) \geq S_{on}(m)\} \leq \sum_{m=0}^{\infty} Pr\{S_{off}(n) \geq S_{on}(m)\} \leq (n+1)^{J^2Q^2} C_2(\Psi_2) 2^{-n\Psi_1}$, where Ψ_1, Ψ_2 are specified in Theorem 3, and*

$$
C_2(\Psi_2) = \sum_{m=0}^{\infty} (m+1)^{J^2Q^2} 2^{-m\Psi_2}. \tag{3.7}
$$

Theorem 4 shows that the probability of exploring a particular distractor path to depth n falls off exponentially with n.

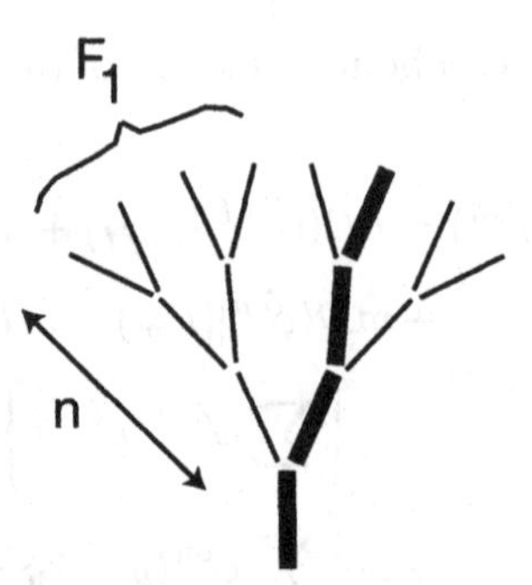

FIG. 9. *This figure illustrates Theorem 5.* F_1 *has at most* Q^n *segments at depth* n.

We now compute the expected number of false segments that are searched at depth n in the tree. There is a factor of $Q^n - 1$ segments at this depth which we can bound above by Q^n.

THEOREM 5. *Provided* $\Psi_1 > \log Q$, *the expected number of segments searched in* F_1 *is less than or equal to* $C_2(\Psi_2)C_1(\Psi_1 - \log Q)$, *where:*

$$C_1(\Psi_1 - \log Q) = \sum_{n=1}^{\infty} n(n+1)^{J^2Q^2} 2^{-n(\Psi_1 - \log Q)}. \tag{3.8}$$

Proof. There are at most Q^n segments at depth n in F_1. Using Theorem 4, the expected number of segments explored is less than or equal to $\sum_{n=0}^{\infty} Q^n (n+1)^{J^2Q^2} C_2(\Psi_2) 2^{-n\Psi_1}$. Sum the series. It will not converge unless $\Psi_1 > \log Q$. □

Finally, by the recursive structure of the tree we see that the number of segments explored in the sets $F_2, F_3, \ldots$ must be less than, or equal to, the number explored in F_1. (We simply eliminate the first few segments, which are in common with the target path, and perform the same argument as above). This yields our main result:

THEOREM 6. *The expected number of segments explored by an A* algorithm using the Bhattacharyya heuristic is bounded above by* $NC_2(\Psi_2)C_1(\Psi_1 - \log Q)$, *provided* $\Psi_1 > \log Q$.

The algorithm is expected to explore $O(N)$ segments and the coefficients $C_2(\Psi_2)C_1(\Psi_1 - \log Q)$ rapidly decrease as Ψ_2 and $\Psi_1 - \log Q$ increase.

In addition, we can estimate the expected error of how much our final estimate differs from the target path. Note that *this is not the same as the error with respect to the MAP estimate.* The error is defined as the number of incorrect segments on the path estimated by A*, see Figure 10.

THEOREM 7. *The expected error is bounded above by* $C_2(\Psi_2)C_1(\Psi_1 - \log Q)$, *provided* $\Psi_1 > \log Q$.

Proof. We measure the error in terms of the expected number of off-road segments. The expected error can be bounded above by $\sum_{n=0}^{\infty} Pr(n)n$, where $Pr(n)$ is the probability that A* will explore a path in F_{N+1-n} to

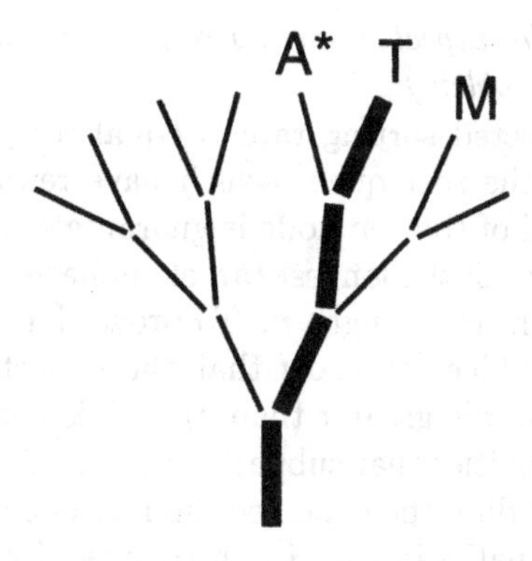

FIG. 10. *Illustrates Theorem 7. A∗ marks the path found by the A* algorithm. M marks the MAP estimate of the target path position. T marks the target path. Observe that A* has an error of one segment and the MAP has an error of two.*

the end. There are Q^n such paths. For each path, the probability that we explore it to the end is bounded above by $\sum_{m=0}^{\infty} Pr\{S_{off}(n) \geq S_{on}(m)\} \leq (n+1)^{J^2Q^2} C_2(\Psi_2) 2^{-n\Psi_1}$, using Theorem 4. Therefore the expected error is bounded above by $\sum_{n=0}^{\infty} nQ^n(n+1)^{J^2Q^2} C_2(\Psi_2) 2^{-n\Psi_1}$, which was summed in Theorem 5. Hence result. □

The condition $\Psi_1 > \log Q$ is related to the order parameter K_B given by equation (2.9). It is straightforward algebra to check that $K_B = \Psi_1 + \Psi_2 - \log Q$. Therefore the condition $\Psi_1 > \log Q$ implies that $K_B > 0$ ($\Psi_2 > 0$ by definition) which ensures that the expected number of paths in F_1 with rewards higher than the target path will fall to zero as $N \mapsto \infty$.

4. Sorting the queue in linear expected time. We have shown that the expected number of nodes searched is linear in N. But the convergence rate of the algorithm will also depend on how much time is required to sort the queue of nodes that we want to expand. In this section, we prove that the expected time to sort the queue nodes is constant.

We use a simple linked list data structure where we order the queue nodes according to their rewards (instead of a more sophisticated data structure, like a heap – see, for example, [5, 3]). A* proceeds by expanding the top node (the one with highest A* reward) and must then adjust the queue to accommodate its children. We now show that the expected sort time, which is required to place the children in their correct positions in the queue, is a constant. To do this, we note that the children nodes have A^* rewards that are smaller than the top node by at most Λ, where $\Lambda = H_L + H_P - \min_y \log P_{on}(y)/P_{off}(y) - \min_t \log P_{\Delta G}(t)/U(t)$ (note that $\Lambda > 0$). We therefore only have to compare the rewards of the children with nodes whose rewards are within Λ of the top node. As we will show, the expected number of these nodes is constant. This gives the following theorem.

THEOREM 8. *The expected sorting rate is constant (i.e. independent of the size N of the problem).*

Proof. The expected sorting rate is equal to Q times the expected number of nodes in the sort queue which have rewards within Λ of the top node. The reward of the top node is guaranteed to be greater than, or equal to, the reward r_T of the longest target subpath in the queue. Let this longest target subpath have length n. To prove that the expected sorting time is constant it suffices to show that the expected number of paths in the queue with rewards greater than $r_T - \Lambda$ is constant. This requires computing the probabilities that subpaths in $F_1, ..., F_n$ have rewards higher than $r_T - \Lambda$ and bounding the expected number of such subpaths. (We do not need to consider paths in $F_i, \quad i > n$ because, by definition of n, they involve children of nodes in the queue and so cannot be in the queue.) We can bound these probabilities using Sanov's theorem and then bound the expected number of nodes by summing exponential series. The details are given in [2]. □

5. Conclusion. The goal of this paper is to point out that Bayesian formulation of inference problems leads to a probability distribution on the ensemble of problem instances. Analysis of this ensemble can give complexity results and, in other work [20, 21], algorithm-independent results.

As a specific example, we analyzed the Geman and Jedynak [6] theory for road tracking. We were able to demonstrate linear (in the road size) expected convergence for a class of ensembles even though the worst case performance is exponential. This agrees with previous work [17] where we analyzed a block-pruning search strategy motivated by [11].

Acknowledgments. We want to acknowledge funding from NSF with award number IRI-9700446, from the Center for Imaging Sciences funded by ARO DAAH049510494, and from the Smith-Kettlewell core grant, and the AFOSR grant F49620-98-1-0197 to A.L.Y. Lei Xu drew our attention to Pearl's book on heuristics and we thank Abracadabra books for obtaining a second hand copy for us. We would also like to thank Dan Snow and Scott Konishi for helpful discussions as the work was progressing and Davi Geiger for providing useful stimulation. David Forsyth, Jitendra Malik, Preeti Verghese, Dan Kersten, Suzanne McKee and Song Chun Zhu gave very useful feedback and encouragement. Finally, we wish to thank Tom Ngo for drawing our attention to the work of Cheeseman and Selman.

REFERENCES

[1] P. CHEESEMAN, B. KANEFSKY, AND W. TAYLOR. "Where the Really Hard Problems are." In *Proc. 12th International Joint Conference on A.I.*. **1**: 331–337. Morgan-Kaufmann. 1991.

[2] J.M. COUGHLAN AND A.L. YUILLE. "Bayesian A* Tree Search with Expected O(N) Node Expansions for Road Tracking." Neural Computation. In press. 2002.

[3] J.M. Coughlan, D. Snow, C. English, and A.L. Yuille. "Efficient Deformable Template Detection and Localization without User Initialization." *Computer Vision and Image Understanding.* **78**: 303–319. June 2000.

[4] T.M. Cover and J.A. Thomas. **Elements of Information Theory**. Wiley Interscience Press. New York. 1991.

[5] D. Geiger and T-L Liu. "Top-Down Recognition and Bottom-Up Integration for Recognizing Articulated Objects." In *Proceedings of the International Workshop on Energy Minimization Methods in Computer Vision and Pattern Recognition.* Eds. M. Pellilo and E. Hancock. Venice, Italy. Springer-Verlag. May. 1997.

[6] D. Geman. and B. Jedynak. "An active testing model for tracking roads in satellite images." *IEEE Trans. Patt. Anal. and Machine Intel.* **18**(1): 1–14. January 1996.

[7] G.R. Grimmett and D.R. Stirzaker. **Probability and Random Processes**. Clarendon Press. Oxford. 1992.

[8] R.M. Karp and J. Pearl. "Searching for an Optimal Path in a Tree with Random Costs." *Artificial Intelligence.* **21**(1,2): 99–116. 1983.

[9] D.C. Knill and W. Richards (Eds). **Perception as Bayesian Inference**. Cambridge University Press. 1996.

[10] S. Konishi, A.L. Yuille, J.M. Coughlan, and S.C. Zhu. "Fundamental Bounds on Edge Detection: An Information Theoretic Evaluation of Different Edge Cues." *Proc. Int'l conf. on Computer Vision and Pattern Recognition*, 1999.

[11] J. Pearl. **Heuristics**. Addison-Wesley. 1984.

[12] B.D. Ripley. **Pattern Recognition and Neural Networks**. Cambridge University Press. 1996.

[13] S. Russell and P. Norvig. "Artificial Intelligence: A Modern Approach. Prentice-Hall. 1995.

[14] B. Selman and S. Kirkpatrick. "Critical Behavior in the Computational Cost of Satisfiability Testing." Artificial Intelligence. **81**(1–2): 273–295. 1996.

[15] V.N. Vapnik. **Statistical Learning Theory**. John Wiley and sons. New York. 1998.

[16] P.H. Winston. **Artificial Intelligence**. Addison-Wesley Publishing Company. Reading, Massachusetts. 1984.

[17] A.L. Yuille and J.M. Coughlan. "Convergence Rates of Algorithms for Visual Search: Detecting Visual Contours." In *Proceedings NIPS'98.* 1998.

[18] A.L. Yuille and J. Coughlan. "An A* perspective on deterministic optimization for deformable templates." *Pattern Recognition.* **33**(4): 603–616. April 2000.

[19] A.L. Yuille and J.M. Coughlan. "Convergence Rates of Algorithms for Visual Search: Detecting Visual Contours." In **Advances in Neural Information Processing Systems 11**. Eds. M.S. Kearns, S.A. Solla, and D.A. Cohn. pp. 641–647. 1999.

[20] A.L. Yuille and J.M. Coughlan. "Fundamental Limits of Bayesian Inference: Order Parameters and Phase Transitions for Road Tracking." *Transactions on Pattern Analysis and Machine Intelligence.* PAMI. **22**: 1–14. February 2000.

[21] A.L. Yuille, J.M. Coughlan, Y-N. Wu, and S.C. Zhu. "Order Parameters for Minimax Entropy Distributions: When does high level knowledge help?" *International Journal of Computer Vision.* **41**(1/2): 9–33. 2001.

EXPECTATION-BASED, MULTI-FOCAL, SACCADIC VISION (UNDERSTANDING DYNAMIC SCENES OBSERVED FROM A MOVING PLATFORM)

ERNST D. DICKMANNS*

Abstract. Based on more than a decade of experience with autonomous road vehicles, a third-generation dynamic vision system has been developed over the last three years. The 'MarVEye' camera configuration realized with three to four conventional miniature CCD-TV-cameras mounted fix relative to each other on a high bandwidth pan-and-tilt platform allows new levels of performance. A saccadic type of vision with intermittent phases of fast saccades and smooth pursuit does complicate image sequence interpretation; however, with explicit spatio-temporal representations available in the 4-D approach this can be handled in a straightforward way. A coherent form of representation of dynamic scenes containing multiple independently moving objects has been achieved through homogeneous coordinate transformations as edges in a scene tree.

The system based on COTS-PC-hard- and software, visual/inertial data fusion and object-oriented programming is presented. Applications discussed are: 1) Turning-off onto a crossroad in a network of minor roads without lane markings, and 2) hybrid adaptive cruise control on high-speed roads with radar/vision data fusion.

1. Introduction. Experience with half a dozen road vehicles equipped with UBM-vision systems has shown that - in the long run - vision systems should have both a wider field of view nearby and higher resolution further away. When looking almost parallel to the plane of motion, the content of images has to be interpreted differently depending on the image line representing different distances in the real world, in general. Homogeneous processing of larger image regions does not make sense because of the different scaling of objects mapped into the image from the real world.

In previous vision systems [Dic 95a] this had led to a bifocal camera arrangement with almost coaxial lines of sight; the fields of view have been about 45 and 15°. For tight maneuvering in curves and for tracking the situation directly in front of the vehicle the former one is too small; for high-speed driving with the requirement of sufficiently high resolution at large look-ahead distances (up to 300m), the latter is too large. In addition, it has been concluded that for conditions where the environment around the points where the vehicle in front touches the ground cannot be seen, monocular range estimation by motion stereo may not be sufficient, e.g. in 'Stop & Go'-traffic or when passing vehicles cut into your lane rather early.

The increased computing power available at low cost, volume and power consumption, now has led to the concept of a '<u>M</u>ulti-focal, <u>a</u>ctive/<u>r</u>eactive <u>V</u>ehicle <u>Eye</u>' (**MarVEye**) built from conventional video components.

*UniBw Munich, LRT/ISF, D-85577 Neubiberg, Germany (Ernst.Dickmanns@unibw-muenchen.de).

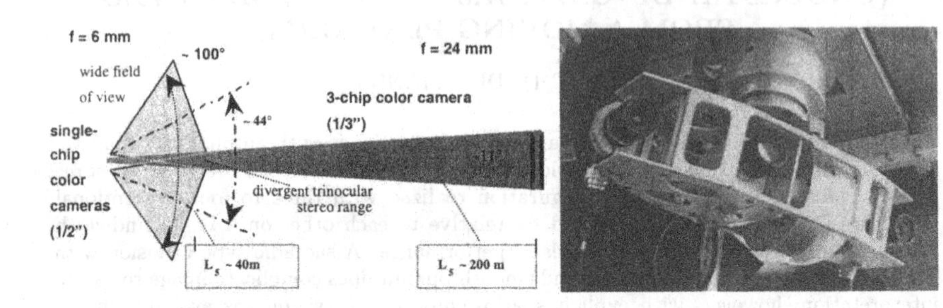

FIG. 1. *MarVEye fields of view and picture of the VaMoRs two-axis pan & tilt platform. The wide angle cameras may be mounted both with parallel optical axes (for stereo vision in a large area) and with divergent axes for a large field of view.*

2. Marveye. Two wide-angle cameras (with a focal length of $f \sim 4$ to 6 mm) may be mounted either with parallel or with divergent optical axes such that there is a central region of overlap providing both a wide simultaneous field of view ($> \sim 100\,°$) and a smaller central area for binocular stereo evaluation. Its separate fields of view are shown in Figure 1. A third camera with a mild tele-lens ($\sim$ 16 to 24 mm) covering part of the stereo region even allows robust trinocular stereo interpretation in its field of view; it is a 3-chip-color-camera with an aperture of about 15 to 23 °.

A fourth camera with an even larger focal length may be added (below the central camera) for good performance in recognition further away (f $\sim$ 50 to 75 mm), allowing a resolution of about 5 cm per pixel at a distance L_5 of $\sim$ 300 to 450 meter. This black-and-white camera may have a high sensitivity with respect to light intensity in order to obtain complementary properties in the camera set; future experience has to show whether a zoom lens on one camera may be preferable.

Each of the images has several hundred-thousand picture elements (pel or pixel). Exploitation of the capabilities thus given requires intelligent viewing direction control. Visually locking onto single moving objects is one of the modes for reduction of motion blur and better performance in recognition of details; the area thus covered is relatively small, however (see areas marked in Fig. 2). If other objects are being noticed in the wide field of view, a fast saccade may be made in order to get one of these objects into better focus. After a short analysis with the higher resolution, further tracking of this object may again be done in the wide-angle images with less total effort. Systematic search patterns are also available for gaze control.

Figure 2 shows three images available for analysis at the same time. The lower two are the skewed wide-angle images, both containing in their

FIG. 2. *Recognition of dirt road from multiple images (MarVEye).*

inner part the same scene in front under slightly different aspect conditions. These two image regions allow binocular stereo interpretation in the near range with a stereo base of about 27 cm. The upper image is the mild tele-image showing the upper central part of the scene in front at much better resolution.

With four cameras, there is a tenfold increase in resolution as compared to the wide-angle cameras. On the other hand, there is a data rate reduction of two orders of magnitude as compared to covering the wide-angle field of view with that high resolution. Of course, intelligent attention focussing has to be introduced in order to make best use of both properties over time.

The saccadic type of vision with intermittent phases of fast saccades and smooth pursuit does complicate image sequence interpretation. However, with explicit spatio - temporal representations available in the 4-D approach this can be handled in a straightforward manner. The internal representation has been adapted to this new multi-focal perception scheme.

3. Joint inertial and visual perception. Initially developed for aircraft applications under gusty conditions, joint inertial and visual perception [Wer 97] with mutual crosswise stabilization has been realized also for road vehicles for handling stronger perturbations in own body motion. Inexpensive angular rate sensors on the moving part of the platform for viewing direction control accurately pick up rotations in a frequency band ranging up to several Hz. The high frequency part of the signal is sufficiently good, while low frequency drifts need special handling for stabilization. Commanding the negative turn rate measured to the viewing direction control leads to high frequency stabilization of the viewing direction to an accuracy of a few tenths of a degree (e.g. 2-axis platform of VaMoRs, see upper left center of Figure 3). This is done at a rate of 500 Hz. On a pitching vehicle, this means that a large part of motion blur induced by this pitching motion will be removed [Scn 95], thereby alleviating image processing and especially range interpretation. In land vehicles, where the gaze direction is almost parallel to the ground, the lines of the stabilized images closely correspond to almost constant look-ahead distances. Slow inertial drifts can easily be stabilized by visual feedback from stationary objects far away (e.g. horizon or some landmark, see Figure 2). Constant scan rates *in inertial space* can thus also be realized despite perturbations.

A set of three mutually orthogonal accelerometers allows picking up accelerations of the own body; a set of angular rate sensors on the vehicle body delivers information on the actual body rotation (including unpredictable perturbation effects). All of these signals are analyzed in order to yield linear velocities and positions or angular orientations (first and second integrals over time). The inertial data rate is 100 Hz at present but may easily be increased if necessary (see upper left in Figure 3). Perspective mapping of other objects into the image planes of various cameras depends on these relative positions (second integrals). Thus, inertial sensing yields lead information for image processing including the effects of unforseeable perturbations. By taking all six (translational and rotational) degrees of freedom into account, short-term predictions can be made without any analytical models for the dynamics of the own rigid body, except the relationships underlying integration over time. Note, that temporal models are unavoidable for data fusion here; the 4-D approach makes use of them right from the beginning. In biological systems like vertebrates, this interaction of inertial and visual sensing is well known from several data paths between vestibular and ocular subsystems.

Because of the huge image data rate in the order of several dozen Mega-Bytes per second (MB/s), data processing has to be done on parallel processors. This incurs time delays of several video cycles, usually, until stable interpretations are achieved in the representation (lower central part of Figure 3). High and low frequency parts of inertial and visual sensor signals are treated and interpreted separately in the overall system. The unavoidable low frequency drift of (low cost) inertial sensors is stabilized by feedback of proper visual signals from objects far away (upper right, Fig. 3).

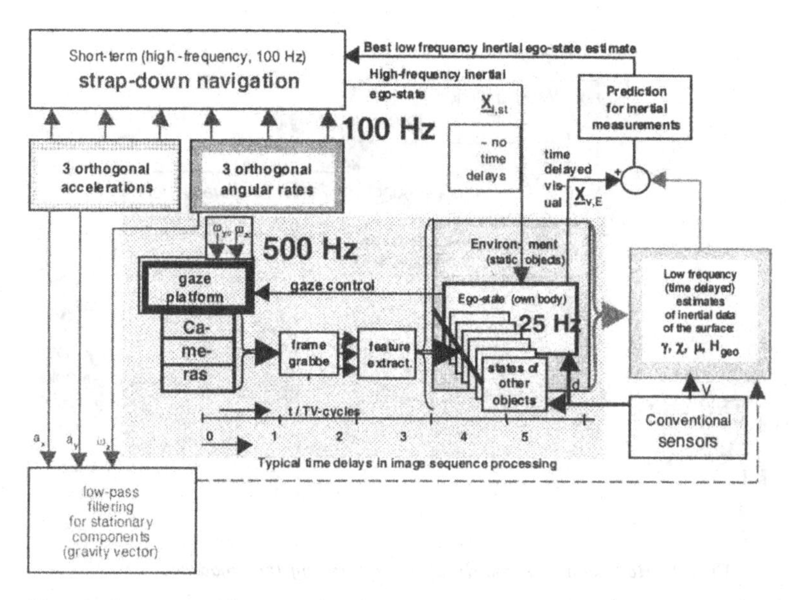

FIG. 3. *Joint visual/inertial (4-D) data interpretation for dynamic ground vehicle guidance.*

4. Knowledge representation. The core representation scheme in the EMS vision system is a stabilized animated world model in 3-D space and time driven in essence by ***inertial and visual*** data. Its central hub is a generic knowledge representation scheme for object and subject classes; subjects are defined as objects with the capability of collecting information and generating control actions based on their own decisions. Differential and integral representations on different scales are used simultaneously [Dic 95b]. Prediction error feedback is the basic method for realizing efficient feature extraction and object tracking in multiple parallel video streams [DiW 99].

Multiple scale representations in 3-D space and time being used span several orders of magnitude. In space, it ranges from pixel size (a few micrometers) to global missions on Earth ($\sim$ 10 000 km) [even planetary lighting conditions by the Sun, 1.5×10^8 km]; in time it ranges from milliseconds for viewing direction control to several hours for mission performance [and years for planetary seasons]. Scaling by a single parameter in homogeneous coordinates (space) or on the time scale is the standard method used. Homogeneous coordinate transformations (HCT) link object positions and orientations in a 'dynamic scene tree' to the image planes in the various cameras [DDi 97]. Figure 4 symbolically shows a road scene with one vehicle in the viewing range of the own vehicle. The mounting conditions for three cameras in this ego-vehicle (lower left) on a gaze control platform are shown in an exploded view on top for clarity (***MarVEye-***

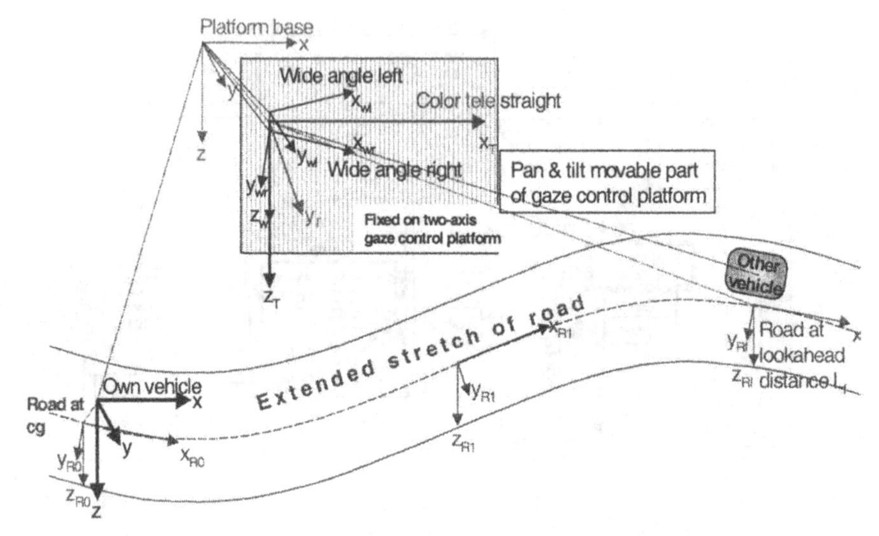

FIG. 4. *Homogeneous coordinates for forming the 'road scene tree'.*

concept). The three coordinate systems, one for each camera, define the different aspect conditions for binocular and trinocular stereo. The trouble in vision, as compared to computer graphics, is that the entries in most of the HCT-matrices are the unknowns of the problem. In a tree representation of this arrangement of objects (see Figure 5 below), each edge between circles represents an HCT and each node (circle) represents an object or sub-object as a movable or functionally separate part. Objects may be inserted or deleted from one frame to the next (dynamic scene tree). The stabilized internal representation has no direct relation to any one of the sensor signals, but is generated taking all sensor signals, their time of measurement and background knowledge into account ('animated world').

Over the last one and a half decades, a systematic approach has been developed for determining the unknown relative state variables by recursive estimation, the so-called '4-D approach' [DiG 88; DiW99]. This makes use of linear approximation matrices for the relationships between feature positions in the various images of the camera set and the state variables or the shape parameters of the models representing the objects seen. These Jacobian matrices have to be determined for each object/sensor pair. They allow recursive least square fits of data to the models thereby bypassing direct perspective inversion (Extended Kalman filtering developed by [Wue 87]). The entries in the Jacobian matrices may either be computed by numerical differencing or by proper storage of intermediate results for the nominal case and multiplication with single derivative matrices.

Figure 5 shows a scene tree with the own body and its essential parts in the lower left (thinly dotted background). In the upper part the road is

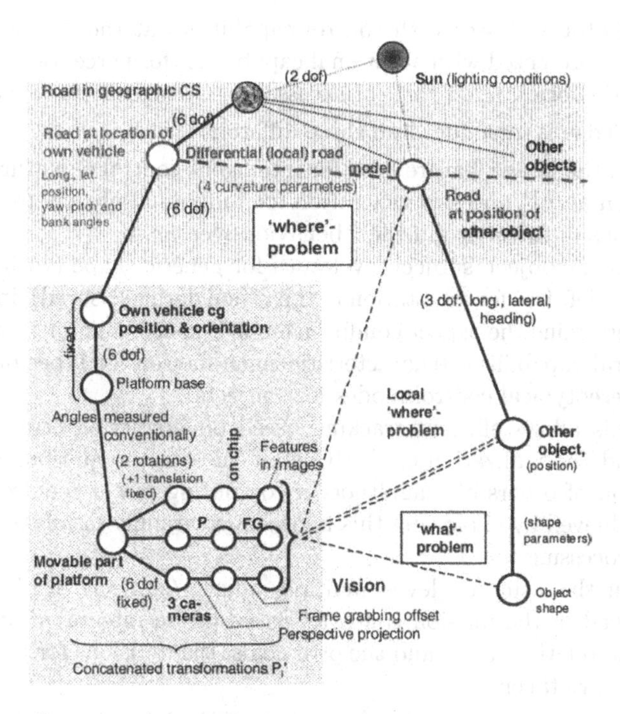

FIG. 5. *Scene tree for visual interpretation of a dynamic scene in EMS-vision showing the individual transformations from the object in 3-D space to the features in the 2-D image stored in computer memory.*

shown, and all six variables describing the relative states between vehicle and road are unknown as well as the geometric parameters of the road in the look-ahead range.

There may be other objects on the road or somewhere relative to the road (e.g. landmarks). Depending on whether one is interested in 'where' the objects are (looking for the center of sums of features) or in 'what' these objects are (usually recognizable by differences between features and their relation over time), two problem areas may be distinguished as known from biological systems as well.

If shadows or the relative direction to the sun are of importance (e.g. driving direction 'into' the sun), their location may be easily represented for each point on the globe and for any point in time by just five more HCT-matrices and the corresponding state variables. In this case the scale factor difference between the pixels in the camera ($\sim 10\ \mu$m) and the distance to the sun (150×10^9 m) is 16 orders of magnitude. (It may be interesting to note that it took humankind thousands of years to work out this simple representation properly.)

Object-oriented programming in C++ has been introduced and nicely fits the generic object classes for the 4-D approach. Subject classes are specialized object classes with control capabilities at their disposal; they have to be represented with additional capabilities for perception, decision-making and control, all of this - if necessary - on various (time) scales.

Knowledge is introduced on three different levels:

1. On the image feature level for intelligent control of feature extraction and hypothesis generation for individual objects from collections of features and their behavior over time.
2. On the object/subject level both for generic shape (with parameters left free for adaptation to the visual data measured), in order to determine the aspect conditions in 3-D space, and for their behavioral capabilities (characteristic eigen-motion for types of objects, stereotypical control modes for subjects).
   This allows efficient tracking based on prediction error feedback and the realization of the 'Gestalt'-idea in recognition. A reduction of orders of magnitude in computing power required can be achieved by exploiting this knowledge carefully in image sequence processing.
3. On the situation level, multiple objects/subjects are being analyzed in the mission context. Based on an integrated analysis of the relative states and the own goals, the decisions for own behavior are taken.

These new scene tree representations with dynamic management of objects have led to much improved flexibility and generality [DDi 97].

Table 1 gives a survey on the different representation scales derived from the 4-D approach as discussed above:

All interaction with the real world takes place 'here and now' (upper left matrix field (1, 1)), especially inertial sensing. Vision is able to also look at the local environment (row 2) yielding features from 'receptive fields' like edges, curvatures and homogeneity of the region. More extended environments (e.g. features moving in conjunction) yield single objects, to which very much of our knowledge about the world is affixed. They are represented by generic classes in row 3. Several objects in conjunction form the geometrical part of the situation and the maneuver space for own actions (row 4). These actions have to be selected in order to fulfil the mission, shown in the last row; there may be several intermediate scales depending on the problem at hand.

Due to the huge data rate of images (tens of MB/s), the second column (temporal differentiation, also called optical flow) is to be avoided. The dynamical models of systems dynamics allow introducing transition matrices for objects to the next point in time when images are going to be taken (third column). Therefore, element (3, 3) is the central hub for image sequence understanding according to the 4-D approach. Since objects move in space and time, maneuvers and mission performance show up in the diagonal from field (3, 3) to the lower right.

TABLE 1

Temporal (horizontal) and 3-D-spatial scales (vertical) as used in EMS-vision for autonomous systems.

range in time → ↓ in space	point in time	temporally local differential environment	local time integrals basic cycle time	extended local time integrals	→	global time integrals
point in space	'here and now' local measurements (1, 1)	temporal change at point 'here' (avoided because of noise amplification)	single step transition matrix derived from notion of (local) 'objects' (row 3)	-------	f e a t	-------
spatially local differential environment	differential geometry: edge angles, positions curvatures	1 " 3a	transition of feature parameters 5	feature history	u r e s	-------
local space integrals → objects	object state, feature- 3b distribution, shape 4	2 motion constraints: diff. eqs., 'dyn. model'	(3, 3) state transition, changed aspect conditions 'central hub'	short range predictions, object state history	o bj e ct s	sparse predictions, object state history
maneuver space of objects	local 3c situation	'lead'-information for efficient controllers	single step prediction of situation (usually not done)	6 multiple step prediction of situation; monitoring of maneuvers	s it u a	-------
↓ . .					ti	
mission space of objects	actual global situation	------	------	monitoring	o n s	7 mission performance, monitoring

5. Behavior control. Decisions for own mission performance are based on local maximization of performance criteria along pre-computed mission plans with certain degrees of flexibility left open for adaptation to the situations actually encountered. Local obstacles are dealt with locally on demand exploiting corresponding capabilities (skills). Both feed-forward and feedback control actuation is used; when feed-forward maneuvers may be subjected to stronger perturbations, a feedback component relative to the nominal case is super-imposed for counteracting the disturbances without intervention from a higher system level. The higher levels have to take care of control mode switching. More details may be found in [Mau 00; Sie 00].

Figure 6 shows the scheme for knowledge representation in the EMS vision system for the situation of turning-off onto a crossroad. Experimen-

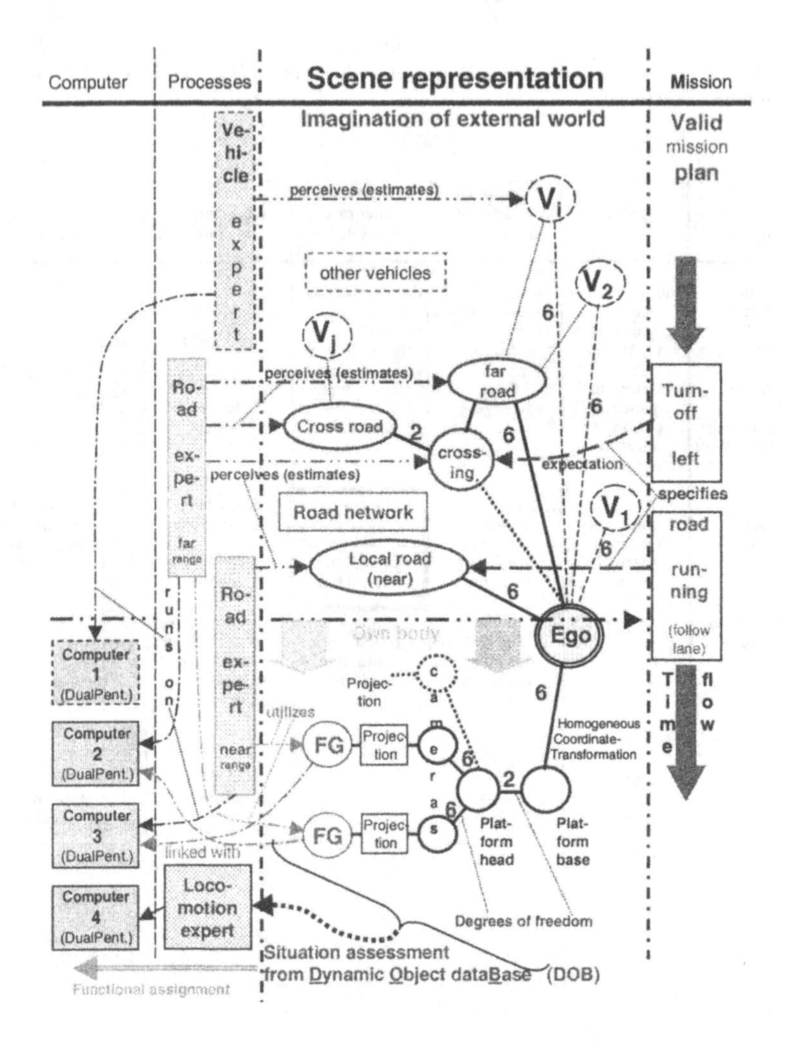

FIG. 6. *Knowledge representation for turning-off onto a crossroad.*

tal results for this application are shown below. This tableau indicates which processes run on which computers (left column), which experts (specialists) are involved (second column), the actual scene tree (in this case the crossroad and the crossing have been inserted) and the actual part of the mission plan (right column). All of this knowledge is evaluated in conjunction. Some aspects of this rather complex control flow and the interdependencies involved will be discussed in the sequel.

This is far beyond computational vision as usually considered in computer science; however, for controlling real-world systems in real-time, this effort has to be made if a flexible overall system with growth potential for

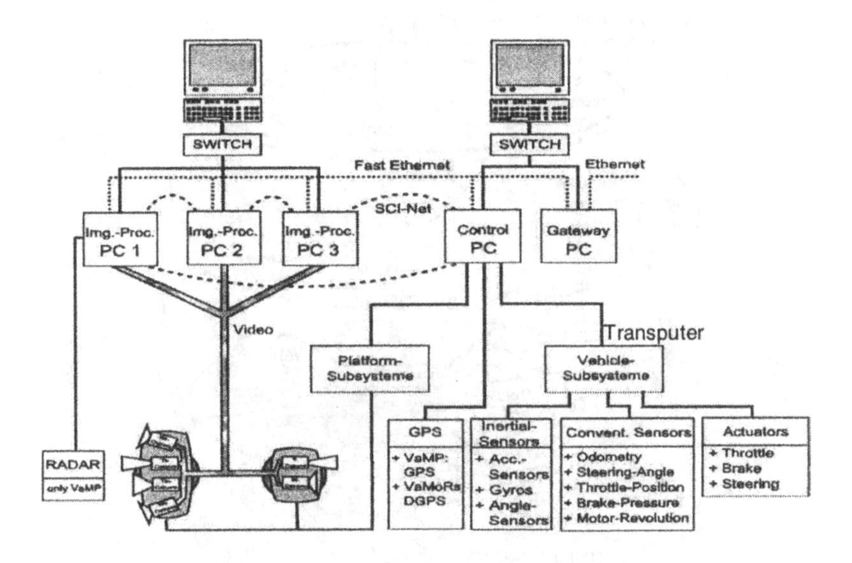

FIG. 7. *Hardware diagram of new COTS-based EMS vision system of UBM.*

future increased computing power is the goal. This goal has been set for the third generation dynamic vision system under discussion.

6. System integration. The system has been realized on commercial-off-the-shelf (COTS) PC-hardware with standard software (four Dual PentiumPro, -/II, -/III with Windows NT, see Figure 7). In parallel, three communication networks are available for different purposes: Ethernet, Fast Ethernet and Scalable Coherent Interface (SCI), the latter one carrying the communication load in real-time operation. All conventional sensors and actuators are linked to a transputer system with four processors for real-time data exchange (called 'vehicle subsystem' in Figure 7; it dates back to the second-generation of vision systems at UBM). System integration takes time delays up to several tenths of a second (for communication and computation from measurement input till actuator output) into account. Specific delay compensations and synchronizations are done for different data paths. Temporal modeling in the 4-D approach provides all the background information needed.

The system is being applied to road vehicle guidance: 1) with the 5-ton van **VaMoRs** for driving on networks of minor roads, and 2) with the Mercedes 500 SEL test vehicle **VaMP** for high speed driving on corresponding roads. This vehicle dates back to the EUREKA-Prometheus program where it was one of the official demonstrators for the company Daimler-Benz in 1994.

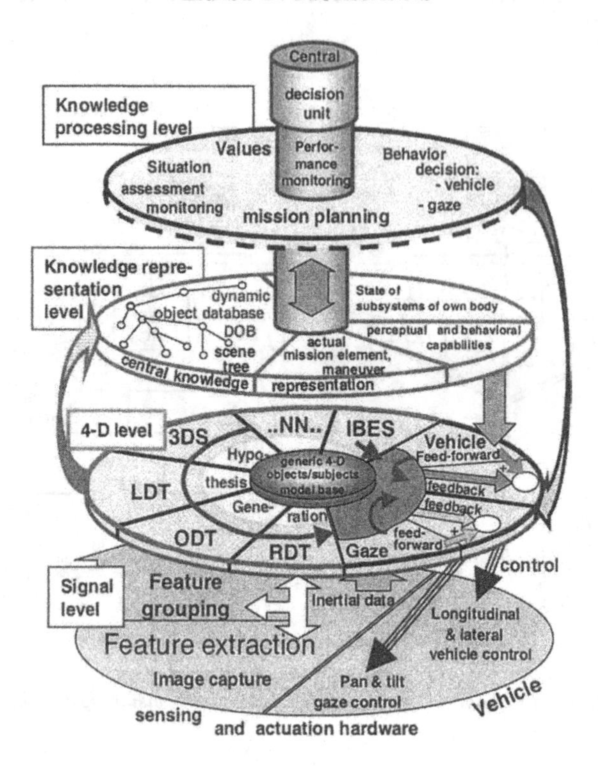

FIG. 8. *Overall cognitive system architecture in EMS-vision (4 layers). RDT = Road detection and tracking; ODT = Obstacle detection and tracking; LDT = Landmark detection and tracking; 3DS = 3D surface recognition; IBES = Inertially-based ego state; NN = Future additional capabilities.*

Beside vision and inertial sensing, both systems have GPS-receivers for determining their global position roughly (selective availability mode); precise localization may be achieved using visual landmark recognition (like road junctions, distinctive road curvatures etc.) in connection with maps [GrD 00].

The overall cognitive architecture of the new EMS-vision system is summarized in Figure 8. This 3-D visualization of the different functional blocks in closed form around a vertical axis on several layers is to symbolize the autonomous capabilities building on each other. The bottom layer is formed by the system hardware for sensing and actuation as well as data (signal) preprocessing without taking temporal models into account. The data rate here is in the tens of MB/s range; feature data rate delivered up to the second (4-D) level is already reduced by one to two orders of magnitude (left hand side of the figure).

It is on this 4-D-level that object oriented spatio-temporal models are used for further reduction of data rates without sacrificing information loss. From collections of features in the image plane, specialists (agents) for detection and tracking of objects of certain classes have to come up with good object hypotheses (shape, aspect conditions, and dynamics) for an efficient interpretation of the motion process observed in 3-D space and time. The dynamical models strictly constrain possible changes from one image to the next, thereby allowing to separate noise from meaningful measurements. Specialists for road detection and tracking (RDT), obstacle detection and tracking (ODT), landmark detection and tracking (LDT), 3-D surface recognition (3DS) and for inertially based (ego-) state estimation (IBSE) are shown. Again, the left side is devoted to bottom-up sensory data input (enriched by information from background knowledge) while the right hand side depicts the top-down decision and control output (see below).

Objects are identified by symbols (ordinal numbers or names), and their relative states and shape parameters are given by a few real numbers representing the entries in the homogeneous coordinate transformations (HCT's) and the generic shape models. In the dynamic object database DOB (see third level, left), objects (or separately movable sub-objects) are represented by circles, while the entries into the HCT's are given by lines. When there is a number written adjacent to the lines they indicate the number of degrees of freedom for this transformation (1 to 6, see also column 3 in Figure 6).

Situation assessment on the fourth (upper, knowledge processing) level observes the evolution of the motion processes for several most interesting other objects and subjects over time in order to find out their possible intentions; it then may predict by fast in-advance simulation whether there is a danger of collision. This requires special actions not directly derivable from actual data alone. This layer also is responsible for implementing the mission plan; it monitors its progress and it has to resolve conflicting requests from specialists for gaze control and attention.

This layer only *takes decisions* with respect to behavior, and it communicates those to the 4-D layer, which is the interface to the actuation hardware. The final implementation runs as hard-real-time process on a controller with direct access to the actuators (see right hand side on lower two levels). The upper two levels have no direct hardware interface; they correspond to purely 'mental' processes based on the internal representations the specialists on level 2 came up with, complemented with their own capability of imagination (simulation with computer graphics type visualization).

7. Experimental results. New experimental results are given here for saccadic viewing direction control with **MarVEye** in **VaMoRs** when approaching an intersection on dirt roads and negotiating a turn-off onto it to the left. Object tracking exploiting MarVEye's capabilities on high-

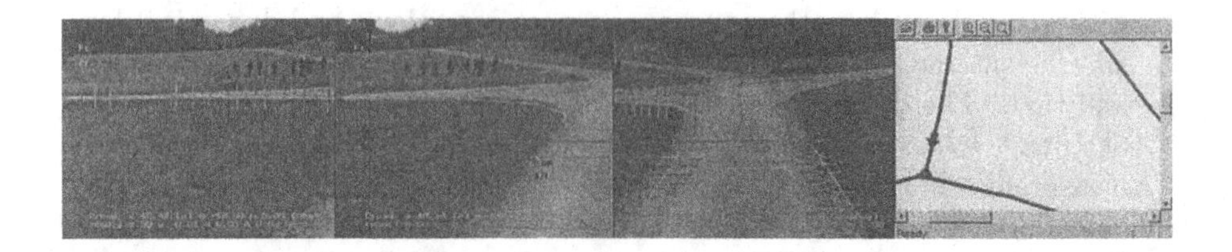

FIG. 9. *Three tele-images from a sequence showing saccadic gaze control while approaching a crossroad (see right sub-image): Looking to the left for detecting the crossroad off the crossing (left sub-image), tracking the own road with the crossing (second from right). During saccadic motion, image evaluation is suppressed (second from left).*

speed roads are demonstrated with **VaMP** in a special application with data fusion from a radar sensor.

Recognition of a crossroad. From Figure 2 the new multi-camera road recognition method with separate models for the near and far range can be seen. In the near range (till about 15 m) covered by the skewed wide-angle cameras, parallel straight lines are fit to edge features and region boundaries on high-speed roads. On minor roads, the full clothoid model may be used. This separation of variables leads to more efficient estimation processes; from this step, road and lane widths as well as the position and orientation of the vehicle relative to the road are obtained. Crossroad detection is first done in the tele-images because of the better spatial resolution. Since their field of view is small, both the intersection with the crossroad and the direction of the crossroad cannot be determined from a single viewing direction.

Therefore, gaze direction is alternated between one point on the crossroad into the turn-off direction and a second one on the intersection. After short periods of fixation on one of these points, a saccade is made to the alternate one until the unknown parameters are reasonably well known. The crossroad is searched in vertical stripes by looking for features indicative of an extended road area. Figure 9 left shows the situation when looking to the left for detecting road features. The changing aspect conditions over time are taken into account by the 4-D approach through measurements from odometry and gaze angles.

Search regions and edge positions discovered are marked in the image. The closed quadrangle marks the actual internal representation for the crossroad; due to a pitching motion of the own vehicle or an error in range estimation a discrepancy may occur. Note that this 4-D interpretation is not a direct inversion of perspective mapping (which could also be done). It is a smoothing recursive estimation process taking the temporal history

FIG. 10. *Wide-angle images right and left after switching to the crossroad as new reference (see text).*

of measurements, its variances, a dynamical model for the ego-motion, and the locomotion of the vehicle measured by odometry into account [Beh 94, Lue 98].

In order to be able to determine the direction of the crossroad relative to the one driven, saccadic eye motions are performed between the intersection (Figure 9 second from right) and a crossroad section to the left. The second image from the left was taken during saccadic motion and is not evaluated.

While the vehicle approaches the intersection, viewing direction is increasingly turned to the left. Figure 10 shows the images from the wide-angle cameras a little while later when the crossroad is already picked up as the new reference. The right road boundary is tracked well in the right image, while on the intersection only a few good features are being found (the left boundary is essentially missing). The left image allows understanding the gaze direction relative to the longitudinal axis of the vehicle. The dark area to the left is the A-frame of the vehicle on the left-hand side; the small side window of **VaMoRs** allows a view on the crossroad far away (brighter curved line). However, image evaluation is done only nearby until the major part of the turn has been performed. A fraction of a second later, the left boundary of the crossroad will be detected in the left image. Based on this information, the look-ahead range is then extended into the region of the tele-lens again, and normal road running is resumed.

Object recognition and tracking. In Figure 11 new results with the hybrid approach using radar and video data in the test vehicle **VaMP** (a Mercedes 500 SEL) are shown. Obstacle detection is done by radar; gaze directions with positive results from the radar signals are marked. The

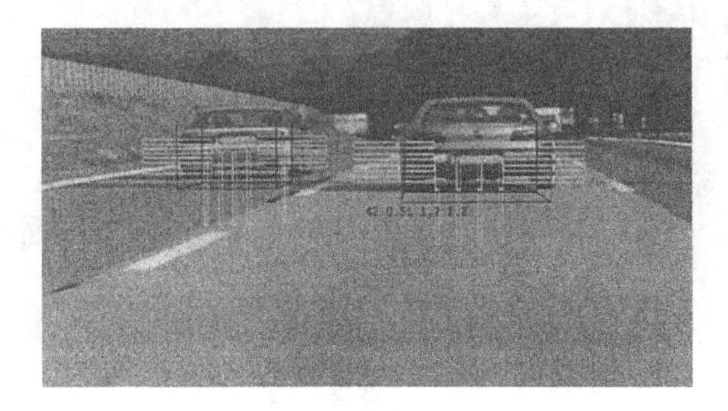

FIG. 11. *Vehicle detection and tracking on a high-speed road.*

range information from radar is good, while azimuth is rather uncertain. Therefore, in the areas surrounding marked directions, well proven visual methods for object tracking [TDD 94] are applied, taking the range information into account. Horizontal and vertical features in conjunction are interpreted as boundaries of vehicles (Figure 11). The visual process yields good information on the lateral position of the vehicle relative to the lanes. The image shown stems from the mild tele-camera and was taken on a high-speed road; the two nearest vehicles are being tracked. For details see [HRD 00].

The newly available processing power will allow area-based image processing (color, various intensity profiles) and much more elaborate analysis of several images in parallel. For example, by directing the strong tele lens onto one of the vehicles in front, image resolution increases by more than a factor of three, which would allow recognizing more details.

Because of the enlarged field of view by the pair of skewed wide-angle cameras, passing vehicles can be detected much earlier. This field of application will receive more attention in the near future for tasks like automatic visual 'Stop & Go' unloading the driver from boring activities.

8. Conclusion and outlook. Dynamic machine vision is going to achieve a new level of performance. Increasing computing power by an order of magnitude every 4–5 years will allow robust and reliable real-world applications not too far into the future. The 4-D approach naturally lends itself for realizing in technical systems the equivalent of vertebrate vision in biology. EMS-vision is considered to be an essential step in this direction. Many characteristic properties of vertebrate vision have been included in this design. Once the approach has demonstrated its superiority over 'minimal' systems with just one camera mounted fix to the body, miniaturized special developments for the vehicle eye may follow.

There is still a long way to go until machine vision systems will come close to performance levels of biological vision systems. However, within the next one or two decades with an increase in computing power per microprocessor of two to four orders of magnitude, and with steady improvements in the knowledge data bases initiated, dramatic changes in dynamic machine vision may be expected.

REFERENCES

[Beh 94] BEHRINGER, R. Road Recognition from Multi-focal Vision. In: I. Masaki (ed.): Proc. of the Internat. Symp. on 'Intelligent Vehicles', Paris, France, 1994, pp. 302–307.

[DDi 97] DICKMANNS, DIRK. Rahmensystem für visuelle Wahrnehmung veränderlicher Szenen durch Computer. Dissertation, UniBwM, Informatik, 1997.

[Dic 95a] DICKMANNS, E.D. Performance Improvements for Autonomous Road Vehicles. Int. Conf.on Intelligent Autonomous Systems (IAS-4), Karlsruhe, 1995.

[Dic 95b] DICKMANNS, E.D. "Parallel Use of Differential and Integral Representations for Realising Efficient Mobile Robots". 7th International Symposium on Robotics Research, Munich, Oct. 1995.

[DiG 88] DICKMANNS, E.D. AND GRAEFE, V. a) Dynamic monocular machine vision. Machine Vision and Applications, Springer International, Vol. 1, 1988, pp. 223–240. b) Applications of dynamic monocular machine vision (ibid), 1988, pp. 241–261.

[DiW 99] DICKMANNS, E.D. AND WÜNSCHE, H.-J. Dynamic vision for Perception and Control of Motion. In Jaehne (ed.): Handbook of Computer Vision and Applications, Vol. 3: Systems and Applications. Acad. Press, 1999, pp. 569–620.

[GrD 00] GREGOR, R. AND DICKMANNS, E.D. EMS-Vision: Mission Performance on Road Networks. Proc. Int. Symp. on Intelligent Vehicles (IV'2000), Dearborn, (MI), Oct. 4–5, 2000.

[HRD 00] HOFMANN, U.; RIEDER, A. AND DICKMANNS, E.D. EMS-Vision: Application to Hybrid Adaptive Cruise Control. Proc. Int. Symp. on Intelligent Vehicles (IV'2000), Dearborn (MI), Oct. 4–5, 2000.

[Lu 98] LÜTZELER, M. AND DICKMANNS, E.D. Robust Road Recognition with MarV-Eye. In: I. Masaki (ed.): Proc. of the Internat. Symp. on 'Intelligent Vehicles', Stuttgart, Germany, Oct. 1998.

[Mau 00] MAURER, M. Flexible Automatisierung von Straßenfahrzeugen mit Rechnersehen. Dissertation, UniBw München, LRT, 21.7.2000.

[Rie 96] RIEDER, A. Trinocular Divergent Stereo Vision. 13th International Conf. on Pattern Recognition, Vienna, Austria, August 1996.

[Scn 95] SCHIEHLEN, J. Kameraplattformen für aktiv sehende Fahrzeuge. Dissertation UniBw München, LRT, 1995.

[Sie 00] SIEDERSBERGER, K.-H. EMS-Vision: Enhanced Abilities for Locomotion. Proc. Int. Symp. on Intelligent Vehicles (IV'2000), Dearborn (MI), Oct. 4–5, 2000.

[TDD 94] THOMANEK, F.; DICKMANNS, E.D. AND DICKMANNS, D. Multiple Object Recognition and Scene Interpretation for Autonomous Road Vehicle Guidance. In: I. Masaki (ed.): Proc. of the Internat. Symp. on 'Intelligent Vehicles', Paris, France, 1994, pp. 231–236.

[Wer 97] WERNER, S. Maschinelle Wahrnehmung für den bordautonomen automatischen Hubschrauberflug. Diessertation, UniBw München, LRT, 17.7.1997.

[Wue 87] WUENSCHE H.-J. Bewegungssteuerung durch Rechnersehen. Dissertation, UniBw München, LRT, 1987.

STATISTICAL SHAPE ANALYSIS IN HIGH-LEVEL VISION

IAN L. DRYDEN*

Abstract. In this article we discuss some of the main aspects of statistical shape analysis and its use in high-level vision. We begin with an introduction to shape and shape space, and then we proceed to describe some common choices of shape metrics. Shape models are discussed in particular detail, and their role as priors in high-level image analysis is described. As an illustration of the methodology we describe an application in Bayesian image analysis for identifying landmarks on face images using a scale-space model.

1. Introduction. Shape is an essential ingredient of high-level image analysis. The geometrical description of an object can be separated into two parts: a) the registration information and b) the 'shape' (which is invariant under registration transformations). A common choice of registration is the group of Euclidean similarity transformations, and then the geometrical properties that are invariant under this group of transformations are known as 'similarity shape'. In a Bayesian approach to object recognition shape information is usually specified as part of the prior distribution. The prior is then combined with the likelihood, or image model, leading to posterior inference about the object.

The statistical theory of shape began with the independent work of Kendall (1977, 1984), Bookstein (1978, 1986) and Ziezold (1977). Subsequent developments have led to a deep differential geometric theory of shape spaces (Kendall et al., 1999), as well as practical statistical approaches to analysing objects using probability distributions of shape and likelihood based inference. A summary of the field is given by Dryden and Mardia (1998), where the main emphasis is on the shapes of labeled point set configurations. In the image analysis literature there are numerous works on the notion of shape, many of which are directly related to the work in Kendall's shape spaces. A common feature of the approaches is some form of shape metric, and many of the shape co-ordinate systems and metrics in common use are approximately linearly related.

In this article we discuss some of the main aspects of statistical shape analysis and its use in high-level vision. In Section 2 we begin with an introduction to shape and shape space, and then we proceed to describe some common choices of shape metrics. In Section 3 shape models are discussed in particular detail, and their role as priors in high-level image analysis is described in Section 4. As an illustration of the methodology we describe an application in Bayesian image analysis for identifying landmarks on face images using a scale-space model.

*School of Mathematical Sciences, University of Nottingham, University Park, Nottingham, NG7 2RD, UK (Ian.Dryden@Nottingham.ac.uk).

2. Shape and shape space.

2.1. General shape. The geometrical description of an object can be decomposed into shape and registration information. The registration information is typically taken as the Euclidean similarity transformations (translation, rotation and isotropic scale). **Shape** is defined to be all the geometrical information which is invariant under registration transformations.

Depending on the application at hand the registration parameters may be of little interest (some applications in shape analysis/morphometrics); the registration and shape may be equally important (object recognition, high-level vision); or the registration parameters are the primary interest (image matching/warping).

Consider the general shape of point sets where k points are available on a manifold M. If the registration group is denoted by G then the general shape space is the orbit space $\Sigma = M^k/G$. The shape distance between $Y, T \in M^k$ is taken to be

$$d_{shape}(Y,T) = \inf_{g \in G} dist(Y, g(T)).$$

We shall primarily focus on the important case of $k \geq m+1$ labeled points in m real dimensions. Let X be a $k \times m$ matrix of the Cartesian co-ordinates of the points (with $M = \mathbb{R}^{km} \setminus$ degeneracy set). Some simple examples are the following:

1. $G =$ Euclidean similarity group (translation, scale and rotation) $= \{\mathbb{R}^m \times \mathbb{R}^+ \times SO(m)\}$
2. $G =$ Isometry group (translation and rotation) $= \{\mathbb{R}^m \times SO(m)\}$
3. $G =$ Affine group (translation, rotation, and shears)

The affine group of transformations is often used in computer vision as an approximation to the projective transformations when the camera is a long way from the object. In this article we shall concentrate on the important example where invariance is due to the Euclidean similarity transformations.

2.2. Two dimensional point sets. When landmarks are available the shape space is quite well understood, especially for 2D point sets (e.g. Kendall, 1984; Bookstein, 1986; Dryden and Mardia, 1998). Consider $k \geq 3$ points in a plane and use complex notation: $z_j \in \mathbb{C}$, $j = 1, \ldots, k$. We remove location (by centering) $z_j - \bar{z}$, $j = 1, \ldots, k$, where $\bar{z} = \frac{1}{k}\sum_{j=1}^{k} z_j$ is the centroid. We then identify scaled and rotated versions as an equivalence class

$$\lambda(z_j - \bar{z}), \quad j = 1, \ldots, k,$$

where $\lambda = re^{i\theta} \in \mathbb{C} \setminus \{0\}$. The shape space is therefore the complex projective space $\mathbb{C}P^{k-2}$ (Kendall, 1984), which is the space of complex lines through the origin (but not including it). So, the challenge from the statistical point of view is to provide models and inferential procedures which are appropriate for the non-Euclidean shape space.

There are several choices of shape distance that could be used and a natural choice is the Procrustes (Riemannian) shape distance between two landmark configurations $z = (z_1, \ldots, z_k)^T$, $w = (w_1, \ldots, w_k)^T$:

$$\rho = \arccos \frac{\sum (z_j - \bar{z})^* (w_j - \bar{w})}{(\sum |z_j - \bar{z}|^2 \sum |w_j - \bar{w}|^2)^{1/2}}, \tag{1}$$

where z^* is the complex conjugate of z^T. The triangle case is special since $\mathbb{C}P^1 \equiv S^2$ (Kendall, 1983). So, the shapes of triangles are represented by points on a sphere and in this case ρ is the great circle distance.

The size of a configuration is often taken to be the centroid size:

$$S(z) = \sqrt{\sum_{j=1}^{k} |z_j - \bar{z}|^2} = \|Cz\|,$$

where $C = I_k - 1_k {1_k}^T / k$ is the $k \times k$ centering matrix, with I_k the $k \times k$ identity matrix and 1_k the k-vector of ones. Other choices such as square root of area could be used, but are not so convenient to work with statistically. In order to represent shape it is often convenient to specify suitable shape co-ordinates, for example Bookstein shape co-ordinates

$$u_j^B = (z_j - z_1)/(z_2 - z_1) - \frac{1}{2}, \quad j = 3, \ldots, k,$$

where u_j^B are the complex co-ordinates of the landmarks after translating, rotating and re-scaling so that point 1 is sent to $-1/2 + 0i$ and point 2 is sent to $1/2 + 0i$. When shape variability is small one can work in a tangent space to shape space, and hence use tangent space co-ordinates (see Dryden and Mardia, 1998, p. 71). For small variations Bookstein co-ordinates and tangent space co-ordinates are approximately linearly related, and hence multivariate normal based inference is approximately equivalent using either shape co-ordinate system (Kent, 1994).

2.3. Higher dimensional point sets. For higher than two dimensional point sets the geometry is not so straightforward. Kendall et al. (1999) discuss the differential geometry of shape spaces in detail, and one particular problem with the higher dimensional shape spaces is that the spaces are not homogeneous and there are singularities.

Nevertheless we can obtain distances and work with care with such higher dimensional spaces. We can consider three steps to obtaining the shape:

1. Remove location (center)

$$X_C = CX.$$

2. Remove size (rescale)

$$Z = \frac{X_C}{S(X)} = \frac{CX}{\|CX\|}$$

which is the preshape which lies on a sphere ($Z \in S^{(k-1)m-1}$).
3. Finally the shape is obtained by identifying all rotated versions as an equivalence class, i.e.

$$[X] = \{Z\Gamma : \Gamma \in SO(m)\},$$

is the shape of X.

So, statistical analysis of shapes can be carried out on the preshape sphere subject to invariance under rotations.

2.4. Shape measures. There are a variety of different choices of shape measures/metrics for shape analysis. Some possible choices are:

1. Procrustes and Riemannian distances
2. Mahalanobis distance in the Procrustes tangent space
3. Sum of all possible angular differences from a template (e.g. Amit, 1997)

and many more.

For the m-dimensional case some specific Procrustes shape distances are:

Partial Procrustes distance:

$$d_P(X_1, X_2) = \inf_{\Gamma \in SO(m)} \|Z_2 - Z_1\Gamma\|.$$

Procrustes/Riemannian metric:

$$\rho(X_1, X_2) = 2 \arcsin(d_P/2), \quad (0 \le \rho \le \pi/2).$$

Full Procrustes distance:

$$d_F(X_1, X_2) = \inf_{r>0,\Gamma} \|Z_2 - rZ_1\Gamma\| = \sin \rho(X_1, X_2).$$

Note that ρ reduces to equation (1) in the two-dimensional case. These distances are all quite similar for shapes which are close together, in particular $d_F = \rho + O(\rho^3) = d_P + O(d_P^3)$.

Sometimes we may have transformations G which are not a group: e.g. a smoothing spline deformation. We can still obtain a discrepancy measure in such cases, e.g. a shape discrepancy measure between $Y, T \in M^k$:

$$D(Y,T) = \inf_{g \in G} dist(Y, g(T))$$

which is not symmetric in T and Y if g is a smoothing thin-plate spline for example. Some comments about non-symmetric shape measures are given by Mumford (1991) and in particular there is empirical evidence from some experiments (Tversky, 1977) that human perception of shape differences between A and B can be different from the difference between B and A, i.e. non-symmetrical shape measures are sometimes required.

In computer vision a common task is to assess how close candidate shapes are to a template, and it is quite possible that observations may be quite variable leading to appreciable differences between different choices of shape measure. However, although different metrics lead to different measures, when comparing two very similar shaped objects the shape distances are often approximately positively linearly related (which can be examined using Taylor series expansions). Hence, often a ranking of the first few closest candidates will be similar due to the approximate positive linear relationship.

Example: Brain templates. We consider an example of template matching where an image is available from an axial MR slice of the brain and the template consists of $k = 8$ landmarks on the corpus callosum (see Amit, 1997, for details). We consider the distance of candidate templates from the standard template in a particular image using three measures listed above at the beginning of the sub-section. In Figure 2 we see pairwise plots of the distances. The general positive correlation is clear and in particular the first candidate template has the same ranking for each distance. However, this example demonstrates that one should be careful in general in the choice of shape measure as the 7th ranked candidate in terms of angular distance is much lower ranked in terms of Riemannian or Mahalanobis distance.

3. Shape variability.

3.1. Shape distributions. There are various approaches for modeling shape variability in objects, for example 1) marginal/offset shape distributions, 2) distributions in preshape space with rotational symmetry, 3) distributions in shape space, 4) distributions in a tangent space. Specifying distributions of shapes is an important component of high-level Bayesian image analysis, where the shape distributions form part of the prior model.

3.2. Marginal/offset distributions. We first consider a model for a configuration in the original space of the landmarks. In particular, we take the mean configuration μ with independent isotropic zero mean normal perturbations with variance σ^2. The marginal or offset normal model is the marginal distribution of shape after integrating out the location, rotation and scale information (see Figure 3).

The offset normal shape density (wrt uniform measure) is (Mardia and Dryden, 1989; Dryden and Mardia, 1991, 1992)

$$\mathcal{L}_{k-2}(-\kappa(1+\cos 2\rho(X,\mu))\exp(-\kappa(1-\cos 2\rho(X,\mu)))$$

where $\kappa = S(\mu)^2/(4\sigma^2)$, $S(\mu)$ is the centroid size of μ and $\mathcal{L}_j(-x) = \sum_{i=0}^{j} \binom{j}{i} \frac{x^i}{i!}$ is the Laguerre polynomial. The parameters are the *Shape*(μ):

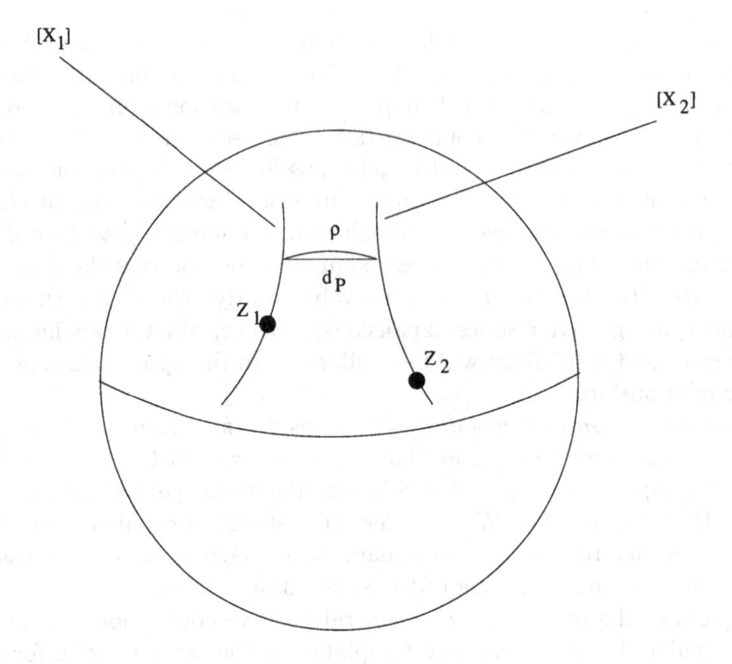

FIG. 1. *A diagrammatic picture of the pre-shape sphere, with the 'curves' (lower dimensional sub-spaces of the pre-shape sphere) represent the shapes corresponding to the preshapes Z_1 and Z_2.*

$2k-4$ mean shape parameters and κ: concentration parameter. For triangles ($k=3$) the shape density is

$$\{1+2\kappa\cos^2\rho(x,\mu)\}\exp\{-2\kappa\sin^2\rho(x,\mu)\}.$$

General covariance matrices and higher dimensions have also be considered Dryden and Mardia (1991) and Goodall and Mardia (1993). The general shape distribution has been used in computer vision by Burl et al. (2001) who use probability ratios based on the shape distribution to decide if candidate configuration shapes are of a similar class to a training set or not.

Inference with marginal shape models can be carried out, such as testing for mean shape difference between two groups (see Dryden and Mardia, 1998, p144, for some examples), although inference is not straightforward for general covariance structures due to over-parameterization.

3.3. Distributions in preshape and shape space. Other shape distributions include the complex Watson distribution (Mardia and Dryden, 1999) with density $f(z)$ proportional to

$$\exp\{-\kappa\sin^2\rho\}$$

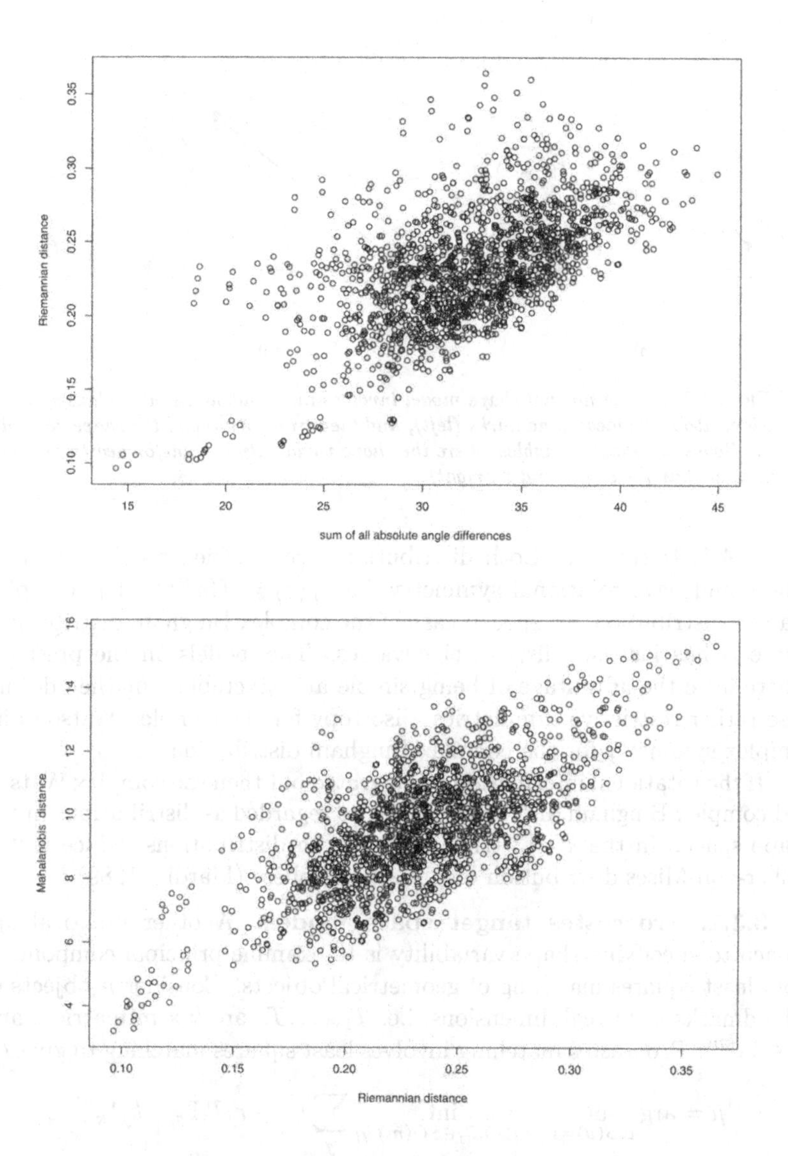

FIG. 2. *Plots of distances to templates in a brain template matching example: (left) Riemannian distance versus sum of all absolute angular differences, (right) Mahalanobis shape distance versus Riemannian distance.*

and the complex Bingham distribution (Kent, 1994) with density $f(z)$ proportional to

$$\exp(z^* Az), \quad \in \mathbb{C}S^{k-1},$$

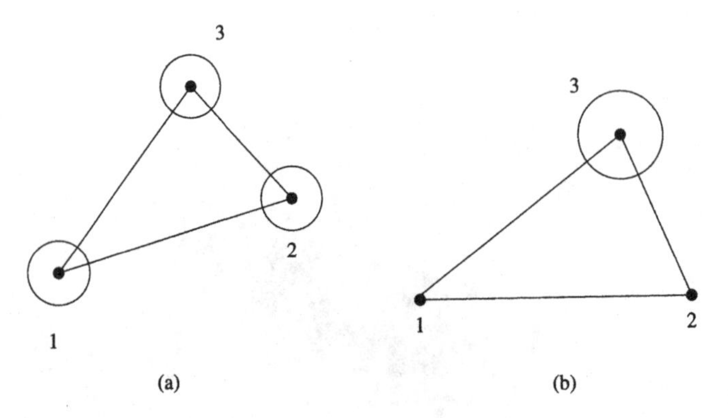

FIG. 3. *The offset normal shape model involves independent circular Gaussian perturbations about the mean landmarks (left), and then transforming to the shape variables such as Bookstein shape variables where the shape variability is transformed to the third vertex after fixing points 1 and 2 (right).*

where A is Hermitian. Both distributions are specified on the preshape sphere and have rotational symmetry, i.e. $f(z) = f(e^{i\theta}z)$. The complex Watson distribution is a special case of the complex Bingham distribution, where A has just two distinct eigenvalues. The models on the preshape sphere have the advantage of being simple and tractable, but they do impose rather restrictive symmetries – isotropy for the complex Watson and complex symmetry for the complex Bingham distribution.

If the rotational information is integrated out then the complex Watson and complex Bingham distributions can be regarded as distributions in the shape space. In the $k = 3$ triangle case both distributions reduce to the Fisher-von-Mises distribution on the shape sphere (Mardia, 1989).

3.3.1. Procrustes tanget space models. Another practical approach to specifying shape variability is to examine principal components from least squares matching of geometrical objects. Consider n objects of k landmarks in m real dimensions, i.e. $T_1, \ldots, T_n$ are $k \times m$ matrices and $T_i \in \mathbb{R}^{km}$. Procrustes matching involves least squares matching to give $\hat{T}_j$:

$$\hat{\mu} = \arg \inf_{\mu: S(\mu)=1} \inf_{r_j>0, \Gamma_j \in SO(m), b_j} \sum_j \|\mu - r_j T_j \Gamma_j + b_j 1_k\|^2$$

where the fitted configurations are $\hat{T}_j = \hat{r}_j T_j \hat{\Gamma}_j + \hat{b}_j 1_k$. Note that if variations are small the shapes lie approximately in a linear space (a tangent space to shape space). Kent and Mardia (2001) give a thorough description of Procrustes tangent space.

After matching the Procrustes mean is $\hat{\mu} = \frac{1}{n} \sum \hat{T}_i$ and the estimated covariance matrix is

$$\hat{\Sigma} = \frac{1}{n} \sum V(\hat{T}_i - \hat{\mu})\{V(\hat{T}_i - \hat{\mu})\}^T$$

where $V(T) = \text{vec}(T)$. For two-dimensional objects the Procrustes matching can be carried out using a complex eigendecomposition (Kent, 1994), but for higher dimensional cases an iterative procedure such as Generalized Procrustes Analysis (Gower, 1975) must be carried out.

The structure of variability in the objects can be examined through the principal components of the Procrustes matched configurations, i.e. through the eigendecomposition of $\hat{\Sigma}$. We can formulate the point distribution model for a two dimensional configuration matrix $X(2k \times 1)$ of k landmarks in $\mathbb{R}^2$ based on the first p PCs as (Mardia, 1997)

$$X = \mu + \sum_{j=1}^{p} y_j \gamma_j + \epsilon, \tag{2}$$

where $y_j \sim N(0, \lambda_j)$, $\epsilon \sim N_{2k}(0, \sigma^2 I)$, independently and the vectors γ_i satisfy

$$\mu^{\mathrm{T}} \gamma_j = 0, \gamma_j^{\mathrm{T}} \gamma_j = 1, \gamma_i^{\mathrm{T}} \gamma_j = 0, \quad i \neq j,$$

and $\lambda_1 \geq \lambda_2 \geq ... \geq \lambda_p$. In addition, for invariance under rotation and for translation, the vectors γ_i satisfy respectively

$$\gamma_j^{\mathrm{T}} \nu = 0 \text{ and } \gamma_j^{\mathrm{T}}(1, \ldots, 1, 0, \ldots, 0)^{\mathrm{T}} = 0, \quad \gamma_j^{\mathrm{T}}(0, \ldots, 0, 1, \ldots, 1)^{\mathrm{T}} = 0,$$

where $\nu = (-\beta_1, \ldots, -\beta_k, \alpha_1, \ldots, \alpha_k)^{\mathrm{T}}$ with $\mu = (\alpha_1, \ldots, \alpha_k, \beta_1, \ldots, \beta_k)^{\mathrm{T}}$. Here $p \leq \min(n-1, 2k-4)$ and p is preferably taken to be quite small, for a parsimonious model.

This method of shape modelling has been used to great success by Cootes et al (1992, 1994) and Kent (1994). Effectively models are specified in the tangent space to the estimated mean, and hence they are appropriate for small variations.

Example: T2 vertebrae In Figure 4 we see an example dataset of 60 landmarks on the outline of T2 mouse vertebrae which have been matched together using Procrustes analysis. The first two principal components of the T2 vertebrae are given in Figure 5. PC1 and PC2 explain 65% and 9% of the shape variability, and PC1 includes the effect of protrusion at the top most part of the bone, and PC2 includes the effect of asymmetry in this part of the bone. Although the interpretation is relatively straightforward here, the PCs can be difficult to interpret in some applications and may consist of multiple effects.

4. High level image analysis and object recognition.

4.1. Deformable templates. An appropriate method for high-level Bayesian image analysis is the use of deformable templates, pioneered by Grenander and colleagues (for example, Grenander, 1994; Grenander and Keenan, 1993). In many applications one has prior knowledge on the composition of the scene and we can formulate parsimonious geometric descriptions for objects in the images. For example, in medical imaging we can

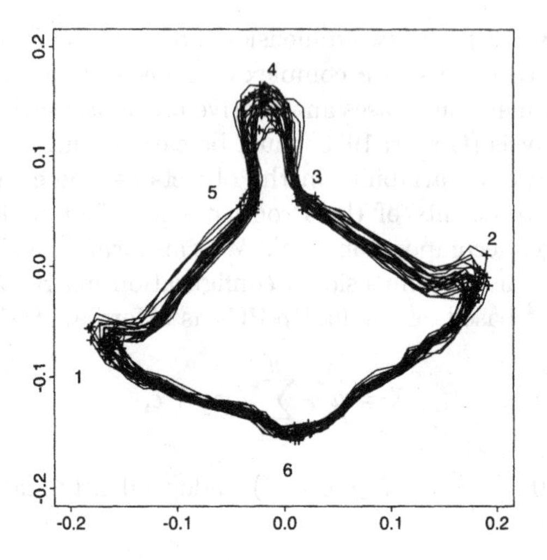

FIG. 4. *A dataset of 23 second thoracic mouse vertebrae that have been matched using Procrustes registration.*

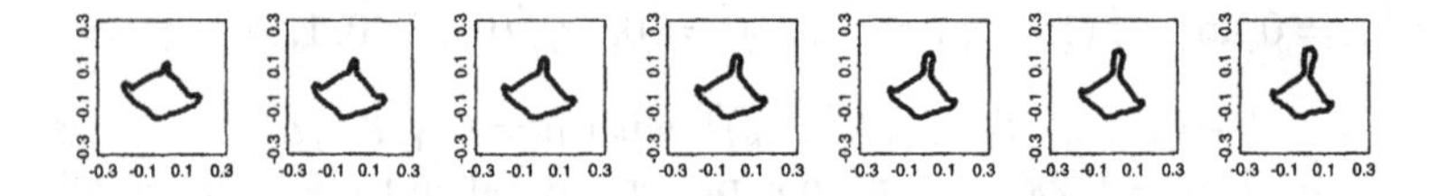

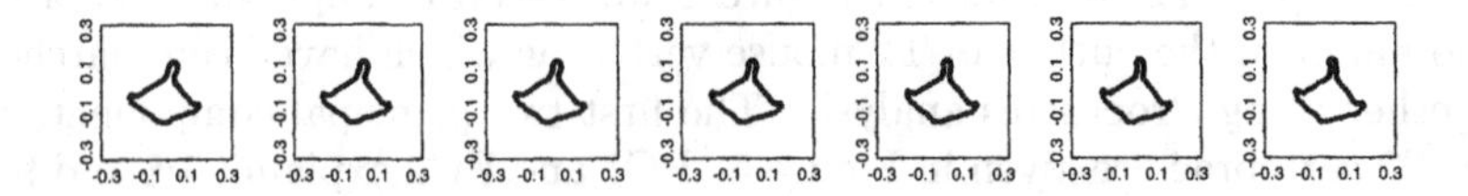

FIG. 5. *The first two principal components of the mouse vertebrae data. The jth row shows PCj, with the ith column displaying $\hat{\mu} + (i-4)\hat{\lambda}_j^{1/2}\hat{\gamma}_j$ where $\hat{\mu}$ is the Procrustes mean, $\hat{\gamma}_j$ is the jth PC and $\hat{\lambda}_j$ is the j eigenvalue of the tangent space covariance matrix.*

expect to know a priori the main subject of the image, e.g. a heart or a brain. Consider our prior knowledge about the objects under study to be represented by a template S_0. Note that S_0 could be a template of a single object or many objects in a scene. A probability distribution is assigned to the parameters with density (or probability function) $\pi(S)$, which models the allowed variations S of S_0. Hence, S is a random vector representing all possible templates with associated density $\pi(S)$. Here S is a function of a finite number of parameters, say $\theta_1, ..., \theta_p$.

In addition to the prior model we require an image model. Let the observed image I be the matrix of grey levels and the image model (or likelihood) is the joint probability density function of the grey levels given the parameterized objects S, written as $L(I|S)$. The likelihood expresses the dependence of the observed image on the deformed template.

By Bayes' Theorem, the posterior density $\pi(S|I)$ of the deformed template S given the observed image I is

$$\pi(S|I) \propto L(I|S)\pi(S). \tag{3}$$

An estimate of the true scene can be obtained from the posterior mode (the maximum a posteriori or MAP estimate) or the posterior mean. The posterior mode is found either by a global search, gradient descent (which is often impracticable due to the large number of parameters) or by techniques such as simulated annealing (Geman and Geman, 1984) or iterative conditional modes (ICM) (Besag, 1986). Alternatively, Markov chain Monte Carlo (MCMC) algorithms (see, for example, Besag et al., 1995; Gilks et al., 1996) provide techniques for simulating from the posterior density.

There is a wide variety of possible template parameterizations that we could consider, including geometrical parameter templates, landmarks/ point distribution models, graphical templates, continuous outline templates, and continuous deformation models (e.g. see Dryden and Mardia, 1998, Chapter 11).

Some possibilities for the image model include i) a scientific model based on the mode of image capture (e.g. Husby et al, 2001), ii) a model based on spatial smoothness (e.g. Gaussian Markov random field), iii) a model based on measurement noise assumptions, iv) a feature density, where particular weight is given to certain features in the image (e.g. McCulloch et al., 1996), or v) combinations of the above. It is often convenient to also include a blurring term in the model.

In the final section we shall consider an application of landmark location on face images. The example involves the use of statistical shape analysis in the construction of a point distribution model, and the image model is based on a scale-space feature density model at each landmark.

4.2. Application. The following application was initially considered by Mardia et al. (1997) who obtained point estimates for the MAP. In the current paper we consider a more complete Bayesian analysis using Markov chain Monte Carlo methods.

4.2.1. Landmarks and features. The aim of the application is to locate a set of reliable landmarks on frontal views of human face images. We first consider an example of such a photograph in Figure 6. A grey level image $I(x, y)$ is available, and we wish to locate landmarks using scale-space features (e.g. McCulloch et al, 1996; Fritsch et al, 1994; Witkin, 1983). A

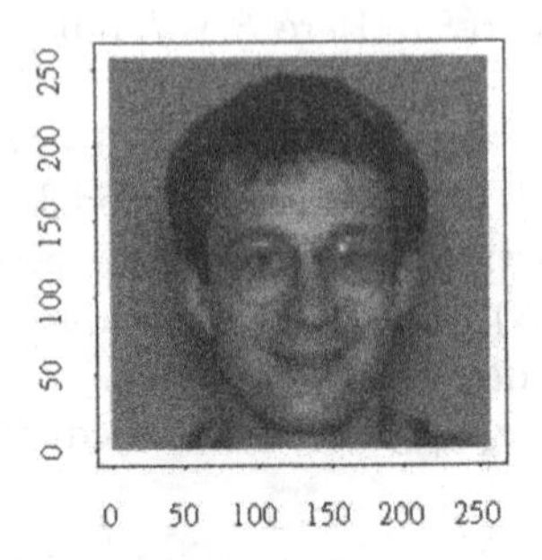

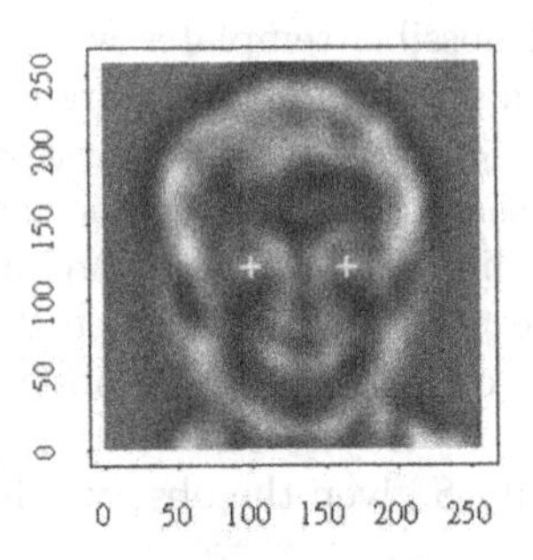

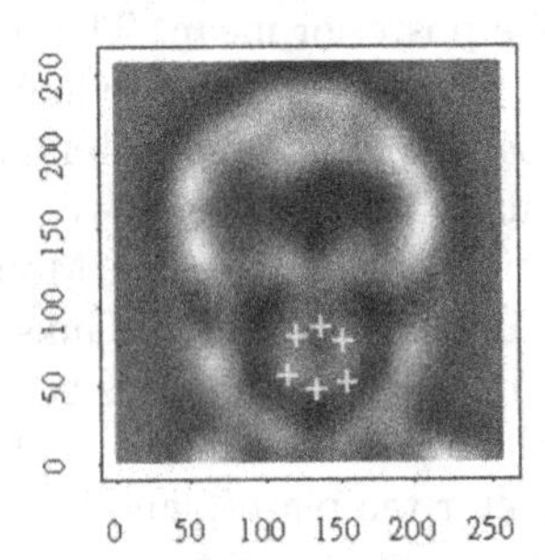

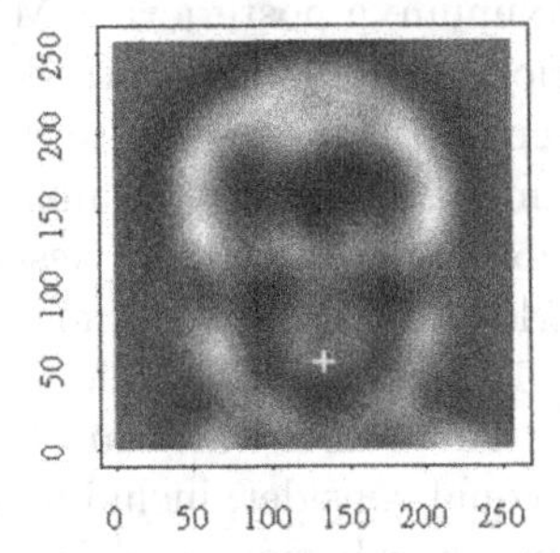

FIG. 6. *An example image displaying the landmarks in scale space: (top left) original image, (top right) two landmarks at scale 8, (bottom left) six landmarks at scale 11, (bottom right) one landmark at scale 13.*

series of scale-space images can be obtained from the original image by a convolution of the image with an isotropic bivariate Gaussian kernel at a succession of 'scales' (σ)

$$S(x, y; \sigma) = \int I(x - h_1, y - h_2) \frac{1}{2\pi\sigma^2} e^{-\frac{1}{2\sigma^2}(h_1^2 + h_2^2)} dh_1 dh_2.$$

The two-dimensional fast Fourier transform is used in the calculation. The particular feature that we are interested in is called the 'medialness' which is the Laplacian of the scale space image:

$$L_{xx}(\sigma) + L_{yy}(\sigma) = \frac{\partial^2 S(x, y; \sigma)}{\partial x^2} + \frac{\partial^2 S(x, y; \sigma)}{\partial y^2}.$$

We describe here a pilot study that was carried out to evaluate the potential of the technique. We initially chose $k = 9$ landmarks (x_i, y_i) on the medialness image at scales 8, 11, 13. The deformable template here is the collection of the nine landmarks, which are displayed in Figure 6.

4.2.2. The image model. The image model that we use is a feature density:

$$L(I|S) \propto \prod_{i=1}^{k} e^{\frac{1}{2}\kappa_i(L_{x_i x_i}+L_{y_i y_i})}.$$

McCulloch et al. (1996) motivate landmark location with this model as mimicking a human observer - focusing on a small narrow area at features and ignoring the rest of the pixels. The features are treated as independent and a high medialness at a feature corresponds to a high density. In effect the non-feature grey levels are treated as independent, uniformly distributed (like a human observer ignoring those pixels). Note that the parameters κ_i need to be specified, and we have fixed them here rather than treating them as hyper-parameters.

4.2.3. The prior model. The image density depends on the parameters of the deformable template (i.e. the landmark locations (x_i, y_i)). In order to model the template effectively we partition the geometrical information into shape information and registration information [location (μ_x, μ_y), rotation θ, and scale $\beta > 0$]. The shape parameters are given by the principal component scores in the tangent space which are obtained from training data. In particular the PC scores c_i are taken to have independent standard normal distributions, and so we are taking the shape distance to be the Mahalanobis distance in the tangent space and are thus using the structural model given in Equation (2).

The prior distributions of the registration parameters are taken to be:

$$\mu_x \sim N(\psi_x, \sigma_x^2),\ \mu_y \sim N(\psi_y, \sigma_y^2)$$
$$\theta \sim N(\psi_t, \sigma_t^2),\ \beta \sim N(\psi_b, \sigma_b^2)$$

where hyperparameters $\psi_x, \sigma_x, \psi_y, \sigma_y, \psi_t, \sigma_t, \psi_b, \sigma_b$ are estimated from training data (10 faces) given in Figure 7. We assume that registration and shape independent and so we have a multivariate normal prior model (configuration density):

$$\pi(S) = \pi(\mu_x, \mu_y, \theta, \beta, c_1, \ldots, c_p)$$
$$\propto \exp\left\{-\frac{1}{2}\left(\frac{(\mu_x-\psi_x)^2}{\sigma_x^2} + \frac{(\mu_y-\psi_y)^2}{\sigma_y^2} + \frac{(\theta-\psi_t)^2}{\sigma_t^2} + \frac{(\beta-\psi_b)^2}{\sigma_b^2} + \sum_{i=1}^{p} c_i^2\right)\right\}$$

Bayes' theorem then gives us the posterior density:

$$\pi(S|I) \propto \pi(S)L(I|S).$$

We carry out statistical inference by drawing samples from the posterior distribution using Markov chain Monte Carlo methods.

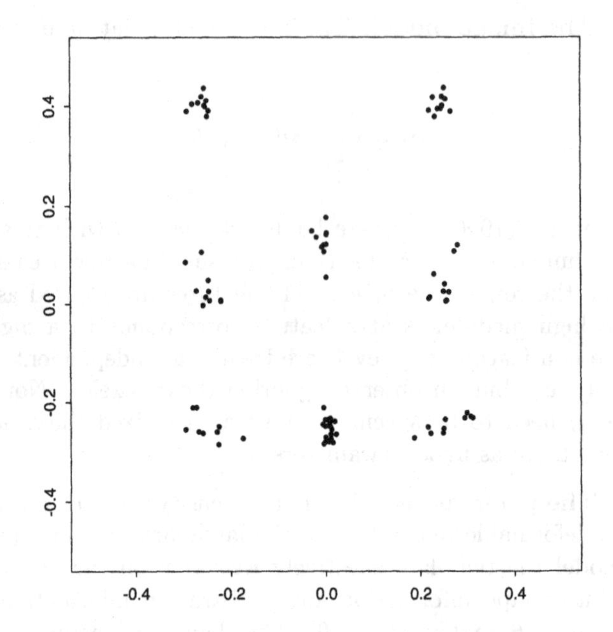

FIG. 7. *The Procrustes rotated face landmarks. There are $n = 10$ configurations of $k = 9$ landmarks.*

In particular the maximum of the posterior or the posterior mean provide suitable estimates of the location of the landmarks, and the associated credibility intervals provide measures of uncertainty. We use a straightforward Metropolis-Hastings algorithm, where the proposal distribution is an independent normal perturbation centred on current observation, with varying variance (linearly decreasing over 5 iterations, then jumping back up). We update each parameter one at a time.

In order to carry out face recognition one could assess which of the candidate templates in the dataset is closest in shape to the estimated landmarks in the current image. We consider briefly some results where the image is face 2 (which is in training set). First of all we plot traces of the posterior, prior and likelihood in Figure 8. We can see that the posterior climbs rapidly over the first few iterations.

In Figure 9 we see time-series plots of the registration parameters and the first two PC scores. Posterior credibility intervals can be obtained for each parameter from these simulated (dependent) samples after removing a burn-in period. In Table 1 we have the 95% prior and posterior credibility intervals for the template parameters. The posterior credibility intervals are obtained using each 10th observation from the final 2000 observations after removing the burn-in period. The MAP estimate is pictured in Figure 10 overlaid on scale 8 and we see that it is close to the hand-picked landmarks used in the training.

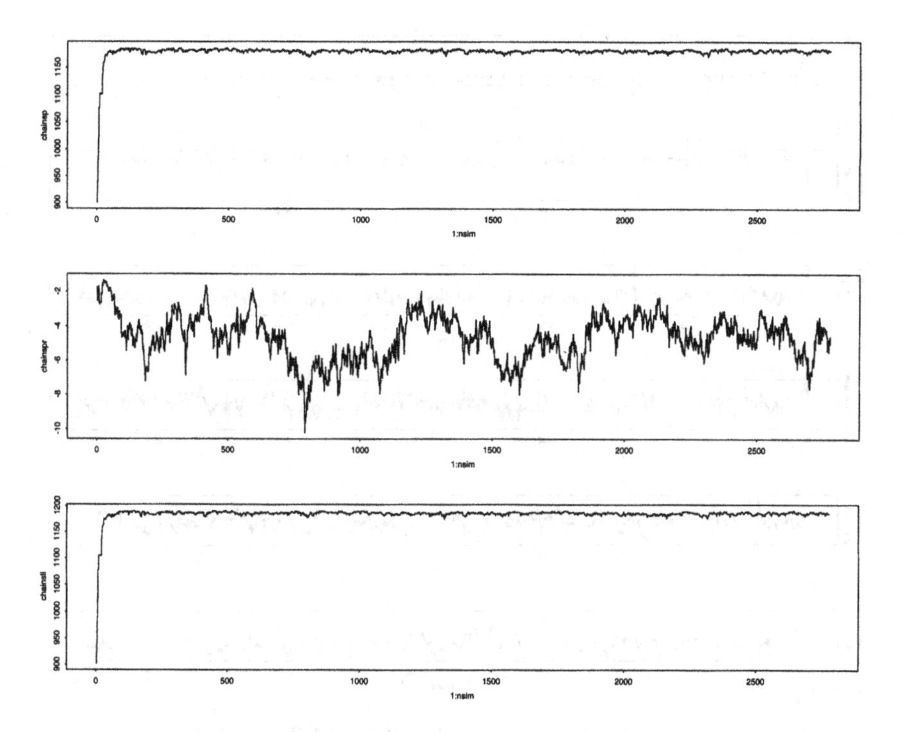

FIG. 8. *Traces of the posterior, prior and likelihood.*

In Figure 11 we see plots of the shape distances from the MAP estimate to each of the 10 faces in the training set. With both choices of Procrustes and Mahalanobis distances the MAP estimate is closest to face 2, which is the correct face for this example. The procedure has worked quite well in other tests, including on images not in the test dataset.

The statistical method described here allows us to give statements about the uncertainty of landmark location. Closely related methods which concentrate on point estimates from scores given to candidate geometric configurations of features include Burl et al. (2001) [using the general shape density], Amit (1997) [using angular shape measures for configurations of feature masks] and Wiskott et al. (1997) [who consider elastic graph matching of features derived from Gabor jets].

5. Discussion. In this article we have described some examples of how statistical shape analysis can be used in high-level vision. There are many other examples of the use of shape in high-level image analysis but we have described the main ingredients of landmark or point set models. See, for example, Loncaric (1998) and Veltkamp (2001) for general reviews of shape measures in pattern recognition.

There is also a wide variety of work on non-landmark shape models, such as snakes (Kass et al., 1988), continuous outline models such as the

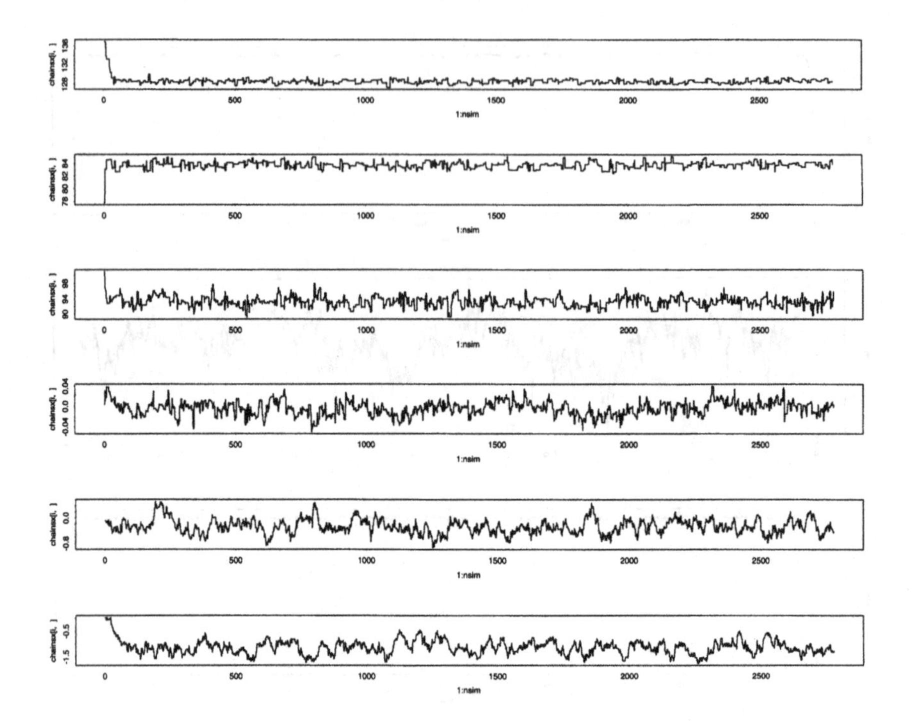

FIG. 9. *Traces of the translations, scale, rotation, PC score 1, PC score 2.*

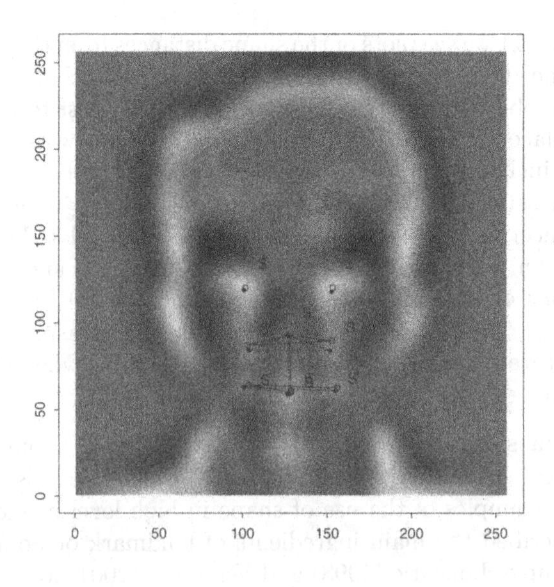

FIG. 10. *The scale space image of the example face at scale 8, with (S) the start point, (black disk) handpicked landmarks, (circle) MAP estimate.*

TABLE 1

Marginal 95% prior and 95% posterior credibility intervals for the registration and shape paramaters for face 2. The posterior credibility intervals are obtained from every 10th observation over the last 2000 iterations of the MCMC scheme.

Parameter	Prior 95% CI	Posterior 95% CI
μ_x	$(119.35, 138.17)$	$(127.98, 129.28)$
μ_y	$(68.76, 97.44)$	$(82.91, 84.58)$
β	$(88.49, 113.66)$	$(90.88, 95.36)$
θ	$(-0.04794, 0.04794)$	$(-0.03093, 0.01636)$
c_1	$(-1.96, 1.96)$	$(-0.622, 0.073)$
c_2	$(-1.96, 1.96)$	$(-1.541, -0.662)$
c_3	$(-1.96, 1.96)$	$(0.893, 2.162)$
c_4	$(-1.96, 1.96)$	$(-2.012, -0.330)$
c_5	$(-1.96, 1.96)$	$(-0.957, 0.727)$
c_6	$(-1.96, 1.96)$	$(-0.135, 1.666)$
c_7	$(-1.96, 1.96)$	$(-1.014, 1.374)$
c_8	$(-1.96, 1.96)$	$(-0.908, 1.665)$

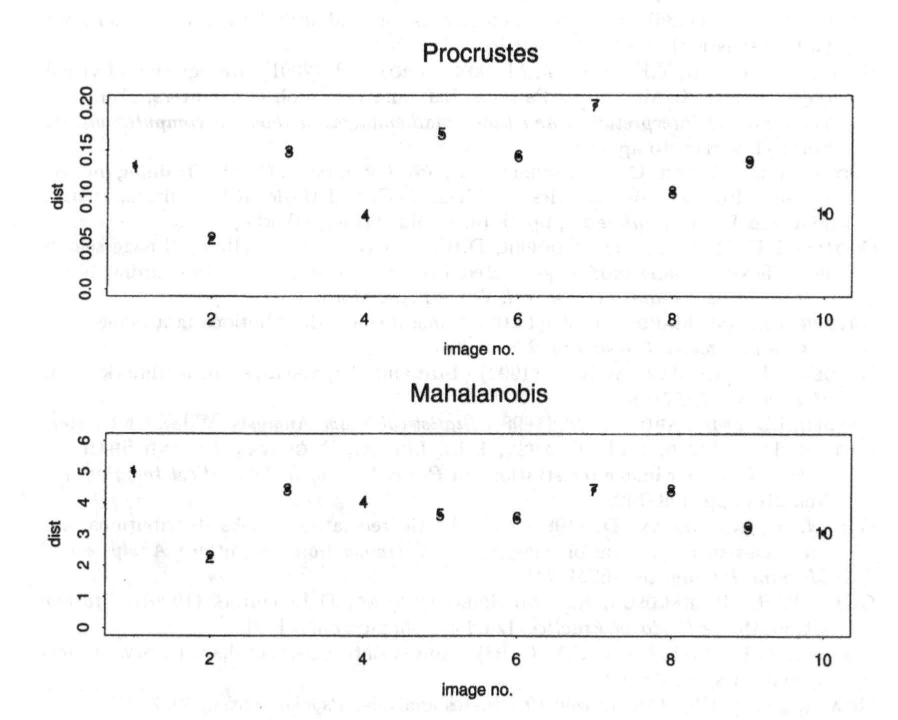

FIG. 11. *Shape distances from the MAP estimate to each of the landmark configurations in the training set: (above) Procrustes distance (below) Mahalanobis distance.*

circular Gaussian Markov random field model (Kent et al., 2000), and Younes' (1998) approach to continuous shape analysis with applications in high-level vision.

An important point to emphasize is that the shape models incorporate high-level knowledge in the inferential procedure. The main advantage of a statistical approach, such as the Bayesian approach described here, is that confidence statements about the parameters can be provided, and these confidence statements are useful in forensic identification and many other applications.

REFERENCES

AMIT, Y. (1997). Graphical shape templates for automatic anatomy detection with applications to MRI scans. *IEEE Transactions on Medical Imaging*, 16:28–40.

BESAG, J., GREEN, P.J., HIGDON, D., AND MENGERSEN, K. (1995). Bayesian computation and stochastic systems. *Statistical Science*, 10:3–66.

BESAG, J.E. (1986). On the statistical analysis of dirty pictures (with discussion). *Journal of the Royal Statistical Society, Series B*, 48:259–302.

BOOKSTEIN, F.L. (1978). *The Measurement of Biological Shape and Shape Change.* Lecture Notes on Biomathematics, Vol. 24. Springer-Verlag, New York.

BOOKSTEIN, F.L. (1986). Size and shape spaces for landmark data in two dimensions (with discussion). *Statistical Science*, 1:181–242.

BURL, M.C., LEUNG, T.K., WEBER, M., AND PERONA, P. (2001). Recognition of visual object classes. In Moons, T., Pauwels, E.J., and van Gool, L.J., editors, *From segmentation to interpretation and back: mathematical methods in computer vision.* Springer-Verlag. to appear.

COOTES, T.F., TAYLOR, C.J., COOPER, D.H., AND GRAHAM, J. (1992). Training models of shape from sets of examples. In Hogg, D.C. and Boyle, R.D., editors, *British Machine Vision Conference*, pp. 9–18, Berlin. Springer-Verlag.

COOTES, T.F., TAYLOR, C.J., COOPER, D.H., AND GRAHAM, J. (1994). Image search using flexible shape models generated from sets of examples. In Mardia, K.V., editor, *Statistics and Images: Vol. 2.* Carfax, Oxford.

DRYDEN, I.L. AND MARDIA, K.V. (1991). General shape distributions in a plane. *Advances in Applied Probability*, 23:259–276.

DRYDEN, I.L. AND MARDIA, K.V. (1992). Size and shape analysis of landmark data. *Biometrika*, 79:57–68.

DRYDEN, I.L. AND MARDIA, K.V. (1998). *Statistical Shape Analysis.* Wiley, Chichester.

FRITSCH, D.S., PIZER, S.M., CHANEY, E.L., LUI, A., RAGHAVAN, S., AND SHAH, T. (1994). Cores for image registration. In *Proceedings of SPIE Medical Imaging '94*, Vol. 2167, pp. 128–142.

GEMAN, S. AND GEMAN, D. (1984). Stochastic relaxation, Gibbs distributions and the Bayesian restoration of images. *IEEE Transactions of Pattern Analysis and Machine Intelligence*, 6:721–741.

GILKS, W.R., RICHARDSON, S., AND SPIEGELHALTER, D.J., editors (1996). *Markov Chain Monte Carlo in Practice*, London. Chapman and Hall.

GOODALL, C.R. AND MARDIA, K.V. (1993). Multivariate aspects of shape theory. *Annals of Statistics*, 21:848–866.

GOWER, J.C. (1975). Generalized Procrustes analysis. *Psychometrika*, 40:33–50.

GRENANDER, U. (1994). *General Pattern Theory.* Clarendon Press, Oxford.

GRENANDER, U. AND KEENEN, D.M. (1993). Towards automated image understanding. In Mardia, K.V. and Kanji, G.K., editors, *Statistics and Images: Vol. 1*, pp. 89–103. Carfax, Oxford.

HUSBY, O., LIE, T., LANGO, T., AND RUE, H. (2001). Bayesian 2-D convolution: a model for diffuse ultrasound scattering. *IEEE Transactions on Ultrasonics, Ferroelectrics and Frequency Control*, 48:121–130.

KASS, M., WITKIN, A., AND TERZOPOULOS, D. (1988). Snakes: active contour models. *International Journal of Computer Vision*, 1:321–331.

KENDALL, D.G. (1977). The diffusion of shape. *Advances in Applied Probability*, 9:428–430.

KENDALL, D.G. (1983). The shape of Poisson-Delaunay triangles. In Demetrescu, M.C. and Iosifescu, M., editors, *Studies in Probability and Related Topics*, pp. 321–330. Nagard, Montreal.

KENDALL, D.G. (1984). Shape manifolds, Procrustean metrics and complex projective spaces. *Bulletin of the London Mathematical Society*, 16:81–121.

KENDALL, D.G., BARDEN, D., CARNE, T.K., AND LE, H. (1999). *Shape and Shape Theory*. Wiley, Chichester.

KENT, J.T. (1994). The complex Bingham distribution and shape analysis. *Journal of the Royal Statistical Society, Series B*, 56:285–299.

KENT, J.T., DRYDEN, I.L., AND ANDERSON, C.R. (2000). Using circulant symmetry to model featureless objects. *Biometrika*, 87(3):527–544.

KENT, J.T. AND MARDIA, K.V. (2001). Shape, tangent projections and bilateral symmetry. *Biometrika*, pp. 469–485.

LONCARIC, S. (1998). A survey of shape analysis techniques. *Pattern Recognition*, 31:983–1001.

MARDIA, K.V. (1989). Shape analysis of triangles through directional techniques. *Journal of the Royal Statistical Society, Series B*, 51:449–458.

MARDIA, K.V. (1997). Bayesian image analysis. *Journal of Theoretical Medicine*, 1:63–77.

MARDIA, K.V. AND DRYDEN, I.L. (1989). Shape distributions for landmark data. *Advances in Applied Probability*, 21:742–755.

MARDIA, K.V. AND DRYDEN, I.L. (1999). The complex Watson distribution and shape analysis. *J. R. Stat. Soc. Ser. B Stat. Methodol.*, 61(4):913–926.

MARDIA, K.V., MCCULLOCH, C., DRYDEN, I.L., AND JOHNSON, V. (1997). Automatic scale-space method of landmark detection. In Mardia, K.V., Gill, C.A., and Aykroyd, R.G., editors, *Proceedings of the Leeds Annual Statistics Research Workshop*, Leeds. University of Leeds Press.

MCCULLOCH, C.C., LAADING, J.K., WILSON, A., AND JOHNSON, V.E. (1996). A shape-based framework for automated image segmentation. In *The American Statistical Association Proceedings of the Section on Bayesian Statistical Science*, pp. 1–6, Chicago, Illinois.

MUMFORD, D. (1991). Mathematical theories of shape: do they model perception? In *Geometric Methods in Computer Vision*, pp. 2–10, Washington. SPIE Proceedings, Vol. 1570.

TVERSKY, A. (1977). Features of similarity. *Psychological Review*, 84:327–352.

VELTKAMP, R.C. (2001). Shape matching: similarity measures and algorithms. Technical report, University of Utrecht, Computer Science. UU-CS-2001-03.

WISKOTT, L., FELLOUS, J.M., KRUEGER, N., AND VON DER MALSBURG, C. (1997). Face recognition by elastic bunch graph matching. *IEEE Transactions on Pattern Analysis and Machine Intelligence*, 19:775–779.

WITKIN, A. (1983). Scale-space filtering. In Bundy, A., editor, *Proceedings of the Eighth International Joint Conference on Artificial Intelligence, Karlsruhe, Germany*, pp. 1019–1022, Los Altos. Kaufman.

YOUNES, L. (1998). Computable elastic distances between shapes. *SIAM J. Appl. Math.*, 58(2):565–586 (electronic).

ZIEZOLD, H. (1977). On expected figures and a strong law of large numbers for random elements in quasi-metric spaces. In *Transactions of the Seventh Prague Conference on Information Theory, Statistical Decision Functions, Random Processes and of the 1974 European Meeting of Statisticians*, Volume A, pp. 591–602, Prague. Academia: Czechoslovak Academy of Sciences.

MAXIMAL ENTROPY FOR RECONSTRUCTION OF BACK PROJECTION IMAGES*

TRYPHON GEORGIOU[†], PETER J. OLVER[‡], AND ALLEN TANNENBAUM[§]

Abstract. Maximum entropy methods have proven to be a powerful tool for reconstructing data from incomplete measurements or in the presence of noise. In this note, we apply the method to the reconstruction computed tomography data derived from backprojection over a finite set of angles. In this case, one derives quite simple formulae which may be easily implemented on computer.

1. Introduction. Computed tomography (CT) is the reconstruction of an image (2D or 3D) from its line of plane integrals. It has been an essential method in diagnostic radiology, and, with the advent of faster scanners of higher resolution, is becoming very important in image guided surgery and therapy as well.

One of the key examples of this technique is when a given cross-section of the body is scanned by an X-ray beam. The intensity loss (which is tissue dependent) is then recorded by a detector, and then computer processed to produce a two dimensional image. There are various possible geometries for the scanners [3], which is important in CT imaging but which we will ignore in this note. The point is that under generic conditions, one can exactly reconstruct a 2D image from its 1D line integrals. The problem is of course that in practice one does not have infinite number of 1D projections, but only a finite number in any given scan. Hence the problem becomes how to find the "best" reconstruction in some suitable sense of the image from such a finite set.

In this note, we propose the use of maximum entropy. Maximum entropy methods have proven to be very important for the reconstruction of data from incomplete measurements or in the presence of noise. For a very nice survey of such results see the paper [1]. Here we apply maximum entropy methods to the problem of reconstruction of images in computed tomography from a finite set of angles. As we will show, in a number of key cases, one can derive some exact formulas for the maximal entropy solution in this framework.

We now summarize the contents of this paper. In Section 2, we outline the theory of the Radon transform, and show how it may be used for image reconstruction. In Section 3, we discuss the methodology of maximal entropy with constraints. Then in Section 4, we give our formulae for

*This work was supported in part by grants from the National Science Foundation, Air Force Office of Scientific Research, and the Army Research Office.

[†]Department of Electrical and Computer Engineering, University of Minnesota, Minneapolis, MN 5545.

[‡]Department of Mathematics, University of Minnesota, Minneapolis, MN 55455.

[§]Depts. of Electrical & Computer and Biomedical Engineering, Georgia Institute of Technology, Atlanta, GA 30332-0250.

optimal reconstruction using a finite set of angles in the maximal entropy sense. Here consider the cases of a continuous density over the image domain, continuous densities for each pixel, and then the discrete case (sampled image and quantized density function). Finally, in Section 5, we sketch some directions for future research on image reconstruction for computerized tomography.

2. The radon transform and image reconstruction. We briefly discuss the Radon transform and its relationship to image reconstruction. The basic problem in CT (computed tomography) is how to reconstruct a 2D image from a set of 1D projections taken along various lines through the image.

More precisely, for $\rho(x, y)$ the intensity map of 2D grey-level image, we consider the integral along a line ℓ_θ which is distance s from the origin and makes an angle θ with the x-axis:

$$g(\theta, s) = \int_{\ell_\theta} \rho(x, y) d\ell_\theta = \int \rho(x, y)\delta(x \sin\theta - y\cos\theta - s)\, dx\, dy. \tag{1}$$

This is precisely the *Radon transform*.

This leads to a simple algorithm for image reconstruction via the so-called *Fourier Slice Theorem.* Namely, it is easy to show that the 1D Fourier transform of function $g(\theta, s)$ is the 2D Fourier transform of the intensity function $\rho(x, y)$. Thus using the the inverse Fourier transform we can reconstruct the image. This leads to a backprojection filter

$$Q(\theta, \omega) = \det J(\omega)\ G(\theta, \omega)$$

where $J(\omega)$ is the Jacobian of the change of coordinates from rectangular to polar, and $G(\theta, \omega)$ is the 2D Fourier transform of $\rho(x, y)$ evaluated at $(\omega \sin\theta, -\omega\cos\theta)$. We then have the filtered backprojection formula:

$$\rho(x, y) = \frac{1}{4\pi^2} \int Q(\theta, \omega) \exp(i\omega(x \sin\theta - y\cos\theta))\, d\omega\, d\theta.$$

Clearly, if one can compute the Radon transform over all angles θ one can reconstruct the image. In practice of course one can only making the computation only a finite sample of angles. The question we now address is what is the "best" backprojection reconstruction over such a finite sample? In the next sections, we give a notion of best reconstruction in information theoretic terms using the notion of maximal entropy.

3. Maximal entropy with constraints. We minimize the functional

$$\int \rho(x, y) \log \rho(x, y)\, dx\, dy, \quad \int \rho(x, y)\, dx\, dy = 1, \quad \rho(x, y) \geq 0, \tag{2}$$

subject to

$$g(\theta, s) = \int_{\ell_\theta} \rho(x,y) d\ell_\theta = \int \rho(x,y)\ \delta(x\cos\theta - y\sin\theta - s)\,dx\,dy.$$

This is equivalent to maximizing the *entropy functional*

$$-\int \rho(x,y)\ \log \rho(x,y)\,dx\,dy.$$

Accordingly we use the method of Lagrange multipliers. Here $\rho(x,y)$ is defined on some subdomain $\Omega \subset \mathbf{R}^2$ (the image domain), which we may take without loss of generality to be $\mathbf{R}^2$.

We now define the integral operator from L^2 to L^2 by

$$A[\rho](\theta, s) := \int \rho(x,y)\ \delta(x\cos\theta - y\sin\theta - s)\,dx\,dy.$$

Notice the with respect to the standard innner product on L^2 we can compute the adjoint operator as follows: For $\lambda(\theta, s) \in L^2$

$$\begin{aligned}\langle A[\rho], \lambda\rangle &= \langle \rho, A^*[\lambda]\rangle \\ &= \int \lambda\left(\theta, s)(\int \lambda(\theta,s)\ \rho(x,y)\ \delta(x\cos\theta - y\sin\theta - s)\,dx\,dy\right)\,d\theta\,ds \\ &= \int \rho(x,y)\ \left(\int \lambda(\theta,s)\ \delta(x\cos\theta - y\sin\theta - s)\,d\theta\,ds\right)\,dx\,dy\end{aligned}$$

from which we see that

$$A^*[\lambda](x,y) := \int \lambda(\theta,s)\ \delta(x\cos\theta - y\sin\theta - s)\,d\theta\,ds.$$

So we introduce the Lagrange multiplier $\lambda(\theta, s)$ and consider the minimization of

$$\int \rho(x,y)\ \log\rho(x,y)\,dx\,dy - \langle\lambda, A[\rho] - g\rangle \tag{3}$$

$$= \int \rho(x,y)\ \log\rho(x,y)\,dx\,dy - \langle\lambda, A[\rho]\rangle - \langle\lambda, g\rangle \tag{4}$$

$$= \int \rho(x,y)\ \log\rho(x,y)\,dx\,dy - \langle A^*[\lambda], \rho\rangle - \langle\lambda, g\rangle. \tag{5}$$

Taking the first variation with respect to ρ yields at a critical point,

$$1 + \log\rho(x,y) = A^*[\lambda](x,y). \tag{6}$$

This is the key relationship for which we will give an explicit solution in the finite dimensional case. Notice that (6) implies that for each $v \in \ker A$,

$$\langle 1 + \log\rho, v\rangle = 0,$$

and so

$$\int \log \rho(x,y) v(x,y)\, dx\, dy = -\int v(x,y)\, dx\, dy.$$

4. Optimal reconstruction using finite set of angles. We consider the problem of best reconstructing an image in the maximal entropy sense using finite set of directions. We work out explicitly the cases for a continous density over the image domain, continuous densities for each pixel, and, finally, the discrete case with sampled image and quantized density function.

4.1. Continuous density over the image domain. We can use the equation (6)

$$1 + \log \rho(x,y) = A^*[\lambda](x,y)$$

to the case of continuous density and finite set of sample angles $\theta_1, \ldots, \theta_n$. In this case, it is easy to see that

$$\begin{aligned} A^*[\lambda](x,y) &= \sum_{i=1}^{n} \int \lambda(\theta_i, s)\delta(x\cos\theta_i - y\sin\theta_i - s)\, ds \\ &= \sum_{i=1}^{n} \lambda(\theta_i, x\cos\theta_i - y\sin\theta_i) = \sum_{i=1}^{n} \lambda_i, \end{aligned}$$

where λ_i is a function supported on the line

$$L_i := \{y\cos\theta_i = -x\sin\theta_i\}, \qquad i = 1, \ldots, n.$$

Thus from equation (6), we see that

$$\rho(x,y) = \prod_i^n a_i, \tag{7}$$

where each a_i is a function whose support is contained in L_i for $i = 1, \ldots n$.

Let us see how this argument looks for horizontal and vertical sections through the image, i.e.,

$$\int \rho(x,y)\, dy = u(x), \qquad \int \rho(x,y)\, dx = v(y).$$

Consistency (Fubini's Theorem) implies

$$\int u(x)\, dx = \int v(y)\, dy = \iint \rho(x,y)\, dx\, dy = s.$$

Our argument above implies that the function $\rho(x,y)$ has to be of "rank one", meaning it is separable

$$\rho(x,y) = a(x)\, b(y).$$

Plugging into the constraint equations says

$$a(x) = \frac{c\,u(x)}{s}, \qquad b(y) = \frac{v(y)}{c},$$

for some constant c, and so

$$\rho(x,y) = \frac{u(x)\,v(y)}{s}$$

is the maximal entropy solution.

4.2. Finite number of continuous pixel density distributions. We are given $\rho_i(x,y) \geq 0$, $i = 1, \ldots N$, pixel density distributions

$$\sum_{i=1}^{N} \rho_i(x,y) = 1.$$

We choose a finite sample of angles

$$\theta = \begin{bmatrix} \theta_1 \\ \vdots \\ \theta_n \end{bmatrix}$$

for distance s to get

$$A \begin{bmatrix} \rho_1 \\ \vdots \\ \rho_N \end{bmatrix} = \begin{bmatrix} g(\theta_1, s) \\ \vdots \\ g(\theta_n, s) \end{bmatrix}. \tag{8}$$

Therefore from equation (6), we see that

$$\begin{bmatrix} 1 + \log \rho_1 \\ \vdots \\ 1 + \log \rho_N \end{bmatrix} = A^* \begin{bmatrix} \lambda_1 \\ \vdots \\ \lambda_n \end{bmatrix}.$$

We have of course $N > n$ (the number of pixels is greater than the number of measurements).

Note that

$$\text{Image of } A^* = (\text{Kernel of } A)^{\perp}.$$

Let $v_1, \ldots, v_r$ be a basis of $\ker A$,

$$v_j = \begin{bmatrix} v_{j1} \\ \vdots \\ v_{jN} \end{bmatrix}, \qquad j = 1, \ldots, r.$$

Therefore

$$v_j \cdot \begin{bmatrix} 1+\log\rho_1 \\ \vdots \\ 1+\log\rho_N \end{bmatrix} = 0, \qquad j=1,\ldots,r,$$

or equivalently

$$\sum_{k=1}^{N} v_{jk}(1+\log\rho_k) = 0, \qquad j=1,\ldots,r.$$

We can compute that

$$-\sum_{k=1}^{N} v_{jk} = \sum_{k=1}^{N} v_{jk}\log\rho_k = \log\prod_{k=1}^{N}\rho_k^{v_{jk}}.$$

Thus we derive the system of equations for ρ_i:

$$\prod_{k=1}^{N}\rho_k^{v_{jk}} = \exp\left(-\sum_{k=1}^{N} v_{jk}\right), \qquad j=1,\ldots,r. \tag{9}$$

We also have the n original (dependent) constraint equations for the densities (8). Note that $\dim \operatorname{Image}(A) = m \leq n$. Then $m+r=N$. From these $n+r$ equations we get $m+r=N$ independent equations for the required densities ρ_i, $i=1,\ldots,N$.

4.3. Discrete density distributions. Let

$$\rho = (\rho_{ij})$$

be an $m \times n$ matrix representing an image. We impose the constraint equations

$$\sum_{j=1}^{n}\rho_{ij} = u_i, \qquad \sum_{i=1}^{m}\rho_{ij} = v_j,$$

corresponding to the row and column sums of ρ. Consistency requires

$$s = \sum_{i=1}^{m} u_i = \sum_{j=1}^{n} v_j = \sum_{i,j=1}^{n}\rho_{ij}.$$

We regard the constraint equations as a linear system $A\rho = g$ of $m+n$ equations in mn unknowns. The elements of the kernel $\ker A$ of the coefficient matrix can be identified with $m \times n$ matrices. A basis for the kernel is given by the matrices v_{ijkl} with $1 \leq i,k \leq m$ and $1 \leq j,l \leq n$, that have two entries equal to $+1$ in positions ij and kl, two entries equal to -1 in positions il and kj, and zeros elsewhere. Therefore, according to Section 4, the maximal entropy solution must satisfy the equations

$$\frac{\rho_{ij}\rho_{kl}}{\rho_{il}\rho_{kj}} = 1$$

or

$$\rho_{ij}\rho_{kl} - \rho_{il}\rho_{kj} = 0$$

for all $1 \leq i, k \leq m$ and $1 \leq j, l \leq n$. The latter system of equations says that all 2×2 minors of the matrix ρ vanish. Therefore

$$\rho = a\, b^T$$

is a matrix of rank 1, i.e.

$$\rho_{ij} = a_i b_j,$$

where $a \in \mathbf{R}^m$ and $b \in \mathbf{R}^n$ are column vectors. Substituting this formula into the constraint equations, we easily find the solution

$$a = \frac{c\,u}{s}, \qquad b = \frac{v}{c}$$

where c is an arbitrary scalar. Therefore

$$\rho_{ij} = \frac{u_i\, v_j}{s}$$

This gives the maximal entropy solution for a general $m \times n$ matrix.

5. Conclusions and further research. In this note, we have begun a rigorous study of the use of a maximal entropy technique for the reconstruction of imagery in computerized tomography. Maximal entropy gives a neat, elegant mathematical solution for this problem.

There are still a number of fundamental issues that must be studied. The first is to describe the procedure for all the various key scanning geometries [2]. This is essential in developing explicit computer algorithms. The next step would be then to actually apply our method to real CT imagery. Robustness to noise artifacts will of course be a major point to be carefully investigated.

REFERENCES

[1] M. Miller and D. Snyder, "The role of likelihood and entropy in incomplete-data problems: applications to estimating point-process intensities and Toeplitz constrained covariances," *Processdings of IEE* **75**: 892–907, 1987.

[2] R. Brooks and G. Di Chiro, "Principles of computer assisted tomography in radiographic and radioisotopic imaging," *Phys. Med. Biol.* **21**: 689–732.

[3] F. Natterer, *The Mathematics of Computerized Tomography*, SIAM Publications, Philadelphia, 2001.
[4] W. Press, S. Teukolsky, W. Vetterling and B. Flannery, *Numerical Recipes in C: The Art of Scientific Computing*, 2d Edition, Cambridge University Press, Cambridge U.K., 1992.
[5] G. Strang, *Introduction to Applied Mathematics*, Wellesley-Cambridge Press, Wellesley, Mass., 1986.

ON THE MONGE–KANTOROVICH PROBLEM AND IMAGE WARPING*

STEVEN HAKER† AND ALLEN TANNENBAUM‡

Abstract. Image registration is the process of establishing a common geometric reference frame between two or more data sets from the same or different imaging modalities possibly taken at different times. In the context of medical imaging and in particular image guided therapy, the registration problem consists of finding automated methods that align multiple data sets with each other and with the patient. In this paper we propose a method of elastic registration based on the Monge–Kantorovich problem of optimal mass transport.

1. Introduction. In this note, we propose a method for image warping and elastic registration based on the classical problem of optimal mass transport. The mass transport problem was first formulated by Monge in 1781, and concerned finding the optimal way, in the sense of minimal transportation cost, of moving a pile of soil from one site to another. This problem was given a modern formulation in the work of Kantorovich [15], and so is now known as the *Monge–Kantorovich problem.*

This type of problem has appeared in econometrics, fluid dynamics, automatic control, transportation, statistical physics, shape optimization, expert systems, and meteorology [21]. It also naturally fits into certain problems in computer vision [9]. In particular, for the general tracking problem, a robust and reliable object and shape recognition system is of major importance. A key way to carry this out is via *template matching,* which is the matching of some object to another within a given catalogue of objects. Typically, the match will not be exact and hence some criterion is necessary to measure the "goodness of fit." For a description of various matching procedures, see [14] and the references therein. The matching criterion can also be considered a *shape metric* for measuring the similarity between two objects.

The registration problem is one of the great challenges that must be addressed in order to make image-guided surgery a practical reality. Registration is the process of establishing a common geometric reference frame between two or more data sets obtained by possibly different imaging modalities. In the context of medical imaging, this is an essential technique for improving preoperative and intraoperative information for diagnosis and image-guided therapy.

*This work was supported in part by grants from the National Science Foundation ECS, NSF-LIS, Air Force Office of Scientific Research, Army Research Office, Coulter Foundation, and MURI Grant.

†Department of Radiology, Surgical Planning Laboratory, Brigham and Women's Hospital, Boston, MA 02115.

‡Departments of Electrical and Computer and Biomedical Engineering, Georgia Institute of Technology, Atlanta, GA 30332-0250.

Indeed, multimodal registration methods play a central role in image-guided therapy systems. First, they allow for the fusing of information from each imaging modality, providing better and more accurate information than can be obtained from each image viewed separately. An example is the fusion of functional imaging with anatomical information from MRI for better localization of damaged brain areas. Second, they allow quantitative comparison of images taken at different times, from which information about evolution over time can be inferred. An example is the monitoring of tumor growth in image sequences. Third, when registering preoperative and intraoperative images, they provide a larger field of view and higher image quality than that available with the intraoperative images alone. An example is the fusion of video images obtained by the laparoscope's video camera with MRI data. Fourth, they allow for the updating of a preoperative image or model using intraoperative tracking data.

Multimodal registration proceeds in several steps. First, each image or data set to be matched should be individually calibrated, corrected for imaging distortions and artifacts, and cleared of noise. Next, a measure of similarity between the data sets must be established, so that one can quantify how close an image is from another after transformations are applied. Such a measure may include the similarity between pixel intensity values, as well as the proximity of predefined image features such as implanted fiducials, anatomical landmarks, surface contours, and ridge lines. Next, the transformation that maximizes the similarity between the transformed images is found. Often this transformation is given as the solution of an optimization problem where the transformations to be considered are constrained to be of a predetermined class. Finally, once an optimal transformation is obtained, it is used to fuse the image data sets.

The method we propose in this paper is designed for elastic registration, and is based on an optimization problem built around the L^2 Kantorovich–Wasserstein distance taken as the similarity measure. The constraint that we will put on the transformations considered is that they obey a mass preservation property. Thus, we will be matching *mass densities* in this method, which may be thought of as weights applied to areas in 2D or volumes in 3D. We will assume that a rigid (non-elastic) registration process has already been applied before applying our scheme.

This type of mass preservation problem occurs naturally in many areas. For example, when registering the proton density based imagery provided by MR. It also occurs in functional MR, where one may want to compare the degree of activity in various features deforming over time, and obtain a corresponding elastic registration map. A special case of this problem occurs in any application where volume or area preserving mappings are considered.

We will give a precise formulation of the problem below (see Section 2), and then develop an algorithm based in part on the work of [10, 17]. The key idea is to find the optimal mapping via an equivalent problem involving certain factorizations (called "polar") of mass preserving mappings. It will turn out that this may be done via a natural gradient descent technique. The details are given in Section 3. In Section 4, we will describe a very general formulation of our framework, and in Section 5 discuss how our ideas may be used for image interpolation and optical flow. We will illustrate our results on some synthetic densities and on real imagery in Section 6.

2. Formulation of the problem. We now give a modern formulation of the Monge–Kantorovich problem. Let Ω_0 and Ω_1 be two subdomains of $\mathbf{R}^d$, with smooth boundaries, each with a positive density function, μ_0 and μ_1, respectively. We assume

$$\int_{\Omega_0} \mu_0 = \int_{\Omega_1} \mu_1$$

so that the same total mass is associated with Ω_0 and Ω_1. We consider diffeomorphisms $\tilde{u}$ from (Ω_0, μ_0) to (Ω_1, μ_1) which map one density to the other in the sense that

$$\mu_0 = |D\tilde{u}|\, \mu_1 \circ \tilde{u} \tag{1}$$

which we will call the *mass preservation* (MP) property, and write $\tilde{u} \in MP$. Equation (1) is called the *Jacobian equation.* Here $|D\tilde{u}|$ denotes the determinant of the Jacobian map $D\tilde{u}$. In words, the Equation (1) means, for example, that if a small region in Ω_0 is mapped to a larger region in Ω_1, then there must be a corresponding decrease in density in order for the mass to be preserved. A mapping $\tilde{u}$ that satisfies this property may thus be thought of as defining a redistribution of a mass of material from one distribution μ_0 to another distribution μ_1.

There may be many such mappings, and we want to pick out an optimal one in some sense. Accordingly, we define the L^p Kantorovich–Wasserstein metric as follows:

$$d_p(\mu_0, \mu_1)^p := \inf_{\tilde{u} \in MP} \int \|\tilde{u}(x) - x\|^p \mu_0(x)\, dx. \tag{2}$$

An *optimal MP map*, when it exists, is one which minimizes this integral. This functional is seen to place a penalty on the distance the map $\tilde{u}$ moves each bit of material, weighted by the material's mass.

The case $p = 2$ has been extensively studied and will the the one proposed in this paper for registration. The L^2 Monge–Kantorovich problem has been studied in statistics, functional analysis, and the atmospheric sciences; see [7, 5] and the references therein. A fundamental theoretical result [16, 6, 11], is that there is a unique optimal $\tilde{u} \in MP$ transporting μ_0 to μ_1, and that this $\tilde{u}$ is characterized as the gradient of a convex function

w, *i.e.* $\tilde{u} = \nabla w$. Note that from Equation (1), we have that w satisfies the *Monge–Ampère* equation

$$|Hw|\, \mu_1 \circ (\nabla w) = \mu_0,$$

where $|Hw|$ denotes the determinant of the Hessian Hw of w.

Hence, the Kantorovich–Wasserstein metric defines a distance between two mass densities, by computing the cheapest way to transport the mass from one domain to the other with respect to the functional given in (2), the optimal transport map in the $p = 2$ case being the gradient of a certain function. The novelty of this result is that like the Riemann mapping theorem in the plane, the procedure singles out a particular map with preferred geometry.

3. Algorithms for computing the transport map. There have been a number of algorithms considered for computing an optimal transport map. For example, methods have been proposed based on linear programming [21], and on Lagrangian mechanics closely related to ideas from the study of fluid dynamics [5]. An interesting geometric method has been formulated by Cullen and Purser [7].

In this section, we will employ a natural solution based on the equivalent problem of *polar factorization*; see [6, 10, 17] and the references therein. We will work with the general case of subdomains in $\mathbf{R}^d$, and point out some simplifications that are possible for the $\mathbf{R}^2$ case.

As above, let Ω_0, $\Omega_1 \subset \mathbf{R}^d$ be subdomains with smooth boundaries, with corresponding positive density functions μ_0 and μ_1 satisfying $\int_{\Omega_0} \mu_0 = \int_{\Omega_1} \mu_1$. Let $u : (\Omega_0, \mu_0) \to (\Omega_1, \mu_1)$ be an initial mapping with the mass preserving (MP) property. Then according to the generalized results of [6, 10], one can write

$$u = (\nabla w) \circ s, \tag{3}$$

where w is a convex function and s is an MP mapping $s : (\Omega_0, \mu_0) \to (\Omega_0, \mu_0)$. This is the *polar factorization* of u with respect to μ_0. In [10], just the case of area preservation is considered, *i.e.* μ_0 is assumed constant, but the general case goes through as well.

Our goal is to find the polar factorization of the MP mapping u, according to the following strategy. We consider the family of MP mappings of the form $\tilde{u} = u \circ s^{-1}$ as s varies over MP mappings from (Ω_0, μ_0) to itself. If we consider $\tilde{u}$ as a vector field, we can always find a function w and another vector field χ, with $\operatorname{div}(\chi) = 0$, such that

$$\tilde{u} = \nabla w + \chi,$$

that is, we can decompose $\tilde{u}$ into the sum of a curl-free and divergence-free vector field [22]. Thus, what we try to do is find a mapping s which will yield a $\tilde{u}$ without any curl, that is, such that $\tilde{u} = \nabla w$. Once such an s is found, we will have $u = \tilde{u} \circ s = (\nabla w) \circ s$ and so we will have found the polar factorization (3) of our given function u.

Now, here is the key point. As we discussed above, the unique optimal solution of the L^2 Monge–Kantorovich problem has the form $\tilde{u} = \nabla w$, and so the problem of finding the polar factorization of u and finding the optimal Monge–Kantorovich mapping $\tilde{u}$ are equivalent. In essence, to solve the Monge–Kantorovich problem we create a "rearrangement" of an initial vector field u using a map s, so that the resulting vector field $\tilde{u} = u \circ s^{-1}$ has no curl. We can now give the technical details.

3.1. Finding an initial mapping. We will now propose an explicit algorithm to solve the Monge–Kantorovich problem. So we want to minimize the L^2 Kantorovich–Wasserstein distance functional over MP functions from (Ω_0, μ_0) to (Ω_1, μ_1). We will try to do this by finding an initial MP mapping u and then minimizing over $\tilde{u} = u \circ s^{-1}$ by varying s over MP mappings from Ω_0 to Ω_0, starting with s equal to the identity map. Our first task is to find and initial MP mapping u. This can be done for general domains using a method of Moser [19, 8], or for simpler domains using the following algorithm. For simplicity, we work in $\mathbf{R}^2$ and assume $\Omega_0 = \Omega_1 = [0,1]^2$, the generalization to higher dimensions being straightforward. We define a function $a = a(x)$ by the equation

$$\int_0^{a(x)} \int_0^1 \mu_1(\eta, y)\, dy\, d\eta = \int_0^x \int_0^1 \mu_0(\eta, y)\, dy\, d\eta \tag{4}$$

which gives by differentiation with respect to x

$$a'(x) \int_0^1 \mu_1(a(x), y)\, dy = \int_0^1 \mu_0(x, y)\, dy. \tag{5}$$

We may now define a function $b = b(x, y)$ by the equation

$$a'(x) \int_0^{b(x,y)} \mu_1(a(x), \rho)\, d\rho = \int_0^y \mu_0(x, \rho)\, d\rho, \tag{6}$$

and set $u(x, y) = (a(x), b(x, y))$. Since $a_y = 0$, $|Du| = a_x b_y$, and differentiating (6) with respect to y we find

$$\begin{aligned} a'(x)\, b_y(x, y)\, \mu_1(a(x), b(x, y)) &= \mu_0(x, y) \\ |Du|\, \mu_1 \circ u &= \mu_0, \end{aligned}$$

which is the MP property we need. This process can be interpreted as the solution of a one-dimensional Monge–Kantorovich problem in the x direction followed by the solution of a family of one-dimensional Monge–Kantorovich problems in the y direction.

3.2. Removing the curl. Once an initial MP u is found, we need to apply the process which will remove its curl. We begin with the following elementary and intuitive property of MP mappings, the proof of which is a simple calculation.

LEMMA. *The composition of two mass preserving (MP) mappings is an MP mapping. The inverse of an MP mapping is an MP mapping.*

Thus, since u is an MP mapping, we have that $\tilde{u} = u \circ s^{-1}$ is an MP mapping if and only if s is, that is, if and only if

$$\mu_0 = |Ds|\,\mu_0 \circ s.$$

In particular, when μ_0 is constant, this equation requires that s be area or volume preserving.

Next, rather than working with s directly, we solve the polar factorization problem via gradient descent. Accordingly, we will assume that s is a function of time, and then determine what $s_t = \frac{\partial}{\partial t} s$ should be to decrease the L^2 Monge–Kantorovich functional. This will give us an evolution equation for s and in turn an equation for $\tilde{u}_t$ as well, the latter being the most important for implementation. By differentiating $\tilde{u} \circ s = u$ with respect to time, we get

$$\begin{aligned}(D\tilde{u} \circ s)\, s_t + \tilde{u}_t \circ s = 0 \\ \tilde{u}_t \circ s = -(D\tilde{u} \circ s)\, s_t \\ \tilde{u}_t = -D\tilde{u}\, s_t \circ s^{-1}.\end{aligned}$$

We need to make sure that s maintains its MP property. Differentiating $\mu_0 = |Ds|\,\mu_0 \circ s$ with respect to time, we derive

$$\begin{aligned}0 &= |Ds|_t\, \mu_0 \circ s + |Ds|\,(\mu_0 \circ s)_t \\ &= |Ds|\left(\operatorname{div}\left(s_t \circ s^{-1}\right) \circ s\right)\, \mu_0 \circ s + |Ds|\,\langle(\nabla\mu_0) \circ s, s_t\rangle, \\ 0 &= \left(\mu_0 \operatorname{div}\left(s_t \circ s^{-1}\right)\right) \circ s\ + \langle(\nabla\mu_0) \circ s, s_t\rangle, \\ 0 &= \mu_0 \operatorname{div}\left(s_t \circ s^{-1}\right) + \left\langle\nabla\mu_0, s_t \circ s^{-1}\right\rangle \\ &= \operatorname{div}(\mu_0\, s_t \circ s^{-1}),\end{aligned}$$

from which we see that s_t and $\tilde{u}_t$ should have the following forms:

$$s_t = \left(\frac{1}{\mu_0}\zeta\right) \circ s, \tag{7}$$

$$\tilde{u}_t = -\frac{1}{\mu_0} D\tilde{u}\,\zeta, \tag{8}$$

for some vector field ζ on Ω_0, with $\operatorname{div}(\zeta) = 0$ and $\langle\zeta, n\rangle = 0$ on $\partial\Omega_0$, n being the normal to the boundary of Ω_0. This last condition ensures that

s remains a mapping from Ω_0 to itself, by preventing the flow of s, given by $s_t = \left(\frac{1}{\mu_0}\zeta\right) \circ s$, from crossing the boundary of Ω_0. This also means that the range of $\tilde{u} = u \circ s^{-1}$ is always $u(\Omega_0) = \Omega_1$.

Consider now the problem of minimizing the Monge–Kantorovich functional:

$$M = \int ||\tilde{u} - i||^2 \mu_0 \tag{9}$$

$$= \int ||\tilde{u}||^2 \mu_0 - 2 \int \langle \tilde{u}, i \rangle \, \mu_0 + \int ||i||^2 \mu_0, \tag{10}$$

where i is the identity map $i(x) = x$. The last term is obviously independent of time. Interestingly, so is the first,

$$\begin{aligned} \int ||\tilde{u}||^2 \mu_0 &= \int ||u \circ s^{-1}||^2 \mu_0 \\ &= \int ||u \circ s^{-1}||^2 \, |Ds^{-1}| \, \mu_0 \circ s^{-1} \\ &= \int ||u||^2 \mu_0 \end{aligned}$$

where $\mu_0 = |Ds^{-1}| \, \mu_0 \circ s^{-1}$ since s^{-1} is an MP map.

Turning now to the middle term, we do a similar trick,

$$\begin{aligned} \int \langle \tilde{u}, i \rangle \, \mu_0 &= \int \langle u \circ s^{-1}, s \circ s^{-1} \rangle \, \mu_0 \\ &= \int \langle u \circ s^{-1}, s \circ s^{-1} \rangle \, |Ds^{-1}| \, \mu_0 \circ s^{-1} \\ &= \int \langle u, s \rangle \, \mu_0, \end{aligned}$$

and taking $s_t = \left(\frac{1}{\mu_0}\zeta\right) \circ s$, we compute

$$\begin{aligned} -\frac{1}{2} M_t &= \int \langle u, s_t \rangle \, \mu_0 \\ &= \int \left\langle \tilde{u} \circ s, \left(\frac{1}{\mu_0}\zeta\right) \circ s \right\rangle \mu_0 \\ &= \int \left\langle \tilde{u} \circ s, \left(\frac{1}{\mu_0}\zeta\right) \circ s \right\rangle |Ds| \, \mu_0 \circ s \\ &= \int \left\langle \tilde{u}, \frac{1}{\mu_0}\zeta \right\rangle \mu_0 \\ &= \int \langle \tilde{u}, \zeta \rangle . \end{aligned}$$

Now decomposing $\tilde{u}$ as $\tilde{u} = \nabla w + \chi$, we have

$$-\frac{1}{2}M_t = \int \langle \nabla w + \chi, \zeta \rangle \tag{11}$$

$$= \int \langle \nabla w, \zeta \rangle + \int \langle \chi, \zeta \rangle \tag{12}$$

$$= \int (\operatorname{div}(w\zeta) - w \operatorname{div}(\zeta)) + \int \langle \chi, \zeta \rangle \tag{13}$$

$$= \int_{\partial\Omega_0} w \langle \zeta, n \rangle + \int \langle \chi, \zeta \rangle \tag{14}$$

$$= \int \langle \chi, \zeta \rangle, \tag{15}$$

where we have used the divergence theorem, $\operatorname{div}(\zeta) = 0$, and $\langle \zeta, n \rangle = 0$ on $\partial\Omega_0$. Thus, in order to decrease M, we can take $\zeta = \chi$ with corresponding formulas (7)–(8) for s_t and $\tilde{u}_t$, provided that we have $\operatorname{div}(\chi) = 0$ and $\langle \chi, n \rangle = 0$ on $\partial\Omega_0$. Thus it remains to show that we can decompose $\tilde{u}$ as $\tilde{u} = \nabla w + \chi$ for such a χ.

Gradient Descent: $\mathbf{R}^d$:

We let w be a solution of the Neumann-type boundary problem

$$\operatorname{div}(\tilde{u}) = \Delta w \tag{16}$$

$$\langle \nabla w, n \rangle = \langle \tilde{u}, n \rangle \text{ on } \partial\Omega_0, \tag{17}$$

and set $\chi = \tilde{u} - \nabla w$. It is then easily seen that χ satisfies the necessary requirements.

Thus, by (8), we have the following evolution equation for $\tilde{u}$:

$$\tilde{u}_t = -\frac{1}{\mu_0} D\tilde{u} \left(\tilde{u} - \nabla \Delta^{-1} \operatorname{div}(\tilde{u}) \right). \tag{18}$$

This is a first order non-local scheme for $\tilde{u}_t$ if we count Δ^{-1} as minus 2 derivatives. Note that this flow is consistent with respect to the Monge–Kantorovich theory in the following sense. If $\tilde{u}$ is optimal, then it is given as $\tilde{u} = \nabla w$, in which case $\tilde{u} - \nabla \Delta^{-1} \operatorname{div}(\tilde{u}) = \nabla w - \nabla \Delta^{-1} \operatorname{div}(\nabla w) = 0$ so that by (18), $\tilde{u}_t = 0$.

Gradient Descent: $\mathbf{R}^2$:

The situation is somewhat simpler in the $\mathbf{R}^2$ case, due to the fact that a divergence free vector field χ can in general be written as $\chi = \nabla^{\perp} h$ for some scalar function h, where $\perp$ represents rotation by ninety degrees, so that $\nabla^{\perp} h = (-h_y, h_x)$. In this case, (15) becomes

$$-\frac{1}{2}M_t = \int \langle \nabla^{\perp} f, \nabla^{\perp} h \rangle = \int \langle \nabla f, \nabla h \rangle \tag{19}$$

where the decomposition of $\tilde{u}$ is $\tilde{u} = \nabla w + \nabla^{\perp} f$, and we can take $h = f$. The function f can be found by solving the Dirichlet-type boundary problem

$$-\operatorname{div}(\tilde{u}^{\perp}) = \Delta f \tag{20}$$

$$f = 0 \text{ on } \partial\Omega_0, \tag{21}$$

which gives us the evolution equation

$$\tilde{u}_t = \frac{1}{\mu_0} D\tilde{u}\, \nabla^{\perp}\Delta^{-1}\operatorname{div}(\tilde{u}^{\perp}). \tag{22}$$

We may also derive a second order *local* evolution equation for $\tilde{u}$ by using the divergence theorem with (19) to get

$$\tilde{u}_t = -\frac{1}{\mu_0} D\tilde{u}\, \nabla^{\perp}\operatorname{div}(\tilde{u}^{\perp}), \tag{23}$$

where appropriate handling of the evolution at the boundary, as described in Section 6, is required. See also our discussion in Section 4.

3.3. Defining the warping map. Typically in elastic registration, one wants to see an explicit warping which smoothly deforms one image into the other. This can easily be done using the solution of the Monge–Kantorovich problem. Thus, we assume now that we have applied our gradient descent process as described above and that it has converged to the Monge–Kantorovich mapping $\tilde{u}_{MK}$.

Following the work of Benamou and Brenier, [5], (see also [11]), we consider the following related problem:

$$\inf \int\int_0^1 \mu(t,x)\|v(t,x)\|^2\, dt\, dx$$

over all time varying densities μ and velocity fields v satisfying

$$\frac{\partial \mu}{\partial t} + \operatorname{div}(\mu v) = 0, \tag{24}$$

$$\mu(0,\cdot) = \mu_0, \quad \mu(1,\cdot) = \mu_1. \tag{25}$$

It is shown in [5] that this infimum is attained for some μ_{min} and v_{min}, and that it is equal to the L^2 Kantorovich–Wasserstein distance between μ_0 and μ_1. Further, the flow $X = X(x,t)$ corresponding to the minimizing velocity field v_{min} via

$$X(x,0) = x, \; X_t = v_{min} \circ X$$

is given simply as

$$X(t,x) = x + t\,(\tilde{u}_{MK}(x) - x). \tag{26}$$

Note that when $t = 0$, X is the identity map and when $t = 1$, it is the solution $\tilde{u}_{MK}$ to the Monge–Kantorovich problem. This analysis provides appropriate justification for using (26) to *define* our continuous warping map X between the densities μ_0 and μ_1. See [18] for applications and a detailed analysis of the properties of this *displacement interpolation*.

4. General case. The Monge–Kantorovich problem can be generalized in the following natural manner. We just sketch the results below. Full details may be found in our joint work with Sigurd Angenent [2].

We first write down the non-local flow. Indeed, suppose we are trying to minimize the following functional over MP maps $\tilde{u}$:

$$M = \int \Phi(\tilde{u} - i)\mu_0 \tag{27}$$

where $\Phi : \mathbf{R}^d \to \mathbf{R}$ is a positive C^1 cost function. In particular, the L^2 Monge–Kantorovich problem above corresponds to the cost function $\Phi(x) = \frac{1}{2}\|x\|^2$.

Since we are still working under a mass preservation constraint, we again define a family of MP maps $\tilde{u} = u \circ s^{-1}$, so that the evolution of $\tilde{u}$ and s is governed by equations (7)–(8):

$$s_t = \left(\frac{1}{\mu_0}\zeta\right) \circ s, \tag{28}$$

$$\tilde{u}_t = -\frac{1}{\mu_0} D\tilde{u}\,\zeta, \tag{29}$$

where ζ is a divergence free vector field. For such a 1-parameter family of maps we have by the change of variables formula and our assumption that $s : \Omega_0 \to \Omega_0$ is mass-preserving,

$$\begin{aligned} M(t) &= \int_{\Omega_0} \Phi(\tilde{u} - i)\mu_0 \\ &= \int_{\Omega_0} \Phi(u - s)\mu_0. \end{aligned}$$

Taking the derivative with respect to time, we find

$$\begin{aligned} M_t &= -\int \langle (\nabla\Phi) \circ (u - s), s_t \rangle\, \mu_0 \\ &= -\int \left\langle (\nabla\Phi) \circ (\tilde{u} - i), \mu_0\ s_t \circ s^{-1} \right\rangle \\ &= -\int \langle \Psi, \zeta \rangle . \end{aligned}$$

where we have used (28) and have defined

$$\Psi := (\nabla\Phi) \circ (\tilde{u} - i). \tag{30}$$

Next, using the Helmholtz decomposition, and imposing boundary conditions for the flow to remain in Ω_0, we take

$$\begin{aligned} \zeta &= \Psi + \nabla p, \\ \mathrm{div}(\zeta) &= 0, \\ \zeta|_{\partial\Omega_0} &\text{ tangential to } \partial\Omega_0. \end{aligned}$$

This leads to the following system of equations:

$$\begin{aligned}\tilde{u}_t &= -\frac{1}{\mu_0} D\tilde{u}(\Psi + \nabla p)\\ \Delta p + \mathrm{div}\Psi &= 0 \text{ on } \partial\Omega_0\\ \frac{\partial p}{\partial \vec{n}} + \langle \Psi, \vec{n}\rangle &= 0 \text{ on } \partial\Omega_0.\end{aligned}$$

This system may also be written in the following nonlocal form, which generalizes (18):

$$\tilde{u}_t = -\frac{1}{\mu_0} D\tilde{u}(\Psi - \nabla\Delta^{-1}\mathrm{div}\Psi). \tag{31}$$

At optimality, the above equations imply that

$$\int \|\Psi + \nabla p\|^2 = 0 \tag{32}$$

so that $\Psi = -\nabla p$, weakly. This is consistent (of course) with known results about the optimal solution of the L^2 Monge–Kantorovich problem. In fact, when an optimum exists (e.g., Φ is positive convex C^1), then this result holds strongly [1].

A purely local flow equation for the generalized Monge–Kantorovich functional may be obtained by setting

$$\zeta = \nabla\mathrm{div}\Psi - \Delta\Psi. \tag{33}$$

It is straightforward to check that in this case $\mathrm{div}(\zeta) = 0$ and

$$M_t = -\frac{1}{2}\int \|\mathrm{curl}\,\Psi\|^2 \le 0. \tag{34}$$

By (29), the corresponding local evolution equation for $\tilde{u}$ is

$$\tilde{u}_t = -\frac{1}{\mu_0} D\tilde{u}\,(\nabla\mathrm{div}\Psi - \Delta\Psi). \tag{35}$$

Notice that at optimality (assuming the solution exists), we must have by (34)

$$\mathrm{curl}\,\Psi = 0,$$

from which we conclude that

$$\Psi = \nabla p,$$

for some function p.

REMARKS. We are also interested in the following type of quadratic function in the Monge–Kantorovich context:

$$\Phi(x, \tilde{u}, D\tilde{u}) := \|\tilde{u} - i\|^2 + \alpha^2 \|D\tilde{u}\|^2.$$

The second term here can be considered as a smoothing term.

This term arises in another context. The concept a harmonic mapping, defined as a minimizer of the Dirichlet integral, can be combined with a mass preservation constraint to obtain a new approach to mass preserving diffeomorphisms [3]. We state the results for Euclidean space even though they apply more generally to Riemannian surfaces. More precisely, let Ω_0, $\Omega_1 \subset \mathbf{R}^d$ be subdomains equipped with positive densities μ_0 and μ_1, respectively, and consider the minimization of the Dirichlet integral over all MP maps:

$$\min_{\tilde{u} \in MP} \int_{\Omega_0} \|D\tilde{u}\|^2. \tag{36}$$

A minimizer, when it exists, is called an MP map of *minimal distortion.*

These maps have applications to brain surface flattening and virtual colonoscopy as described in [4, 13]. The numerical procedure and applications will be presented in a future paper.

5. Image interpolation. In this section, we indicate how one can employ optimal mass transport ideas in order to formulate a new method for image interpolation, and as a natural corollary for the computation of optical flow. (See also [23] and the references therein.) The idea is to minimize a functional of the following form over area-preserving maps:

$$M = \int (F \circ \tilde{u} - G)^2 + \alpha^2 \|\tilde{u} - i\|^2. \tag{37}$$

Here the first term controls the "goodness of fit" between the (intensity) images $G : \Omega_0 \to \mathbf{R}$ and $F : \Omega_1 \to \mathbf{R}$, and the second Monge–Kantorovich term controls the the warping of the map. In this term, we are trying to penalize the distance of $\tilde{u}$ from the identity. As with the Monge–Kantorovich flows, we can assume $\tilde{u} = u \circ s^{-1}$ where u is an initial mapping from Ω_0 to Ω_1, and s is a family of area-preserving mappings from Ω_0 to itself parameterized by "time" in order to compute the first variation.

More generally, one may consider the following type of problem:

$$\text{Goodness of fit} + \alpha^2 \int \|\tilde{u} - i\|^2.$$

In equation (37), we took

$$\text{Goodness of fit} = \int (F \circ \tilde{u} - G)^2.$$

But one can choose a term based on the concept of *mutual information,* for example as in [24]. Note that our approach gives natural correspondence measures for a sequence of images (e.g., as in a video), and thus can be tailored for use in optical flow and tracking. Further, the methodology given above for optimal transport can be used in an analogous manner to derive local and non-local schemes for decreasing the functional (37).

6. Implementation and examples. We illustrate our methods with two examples. The first is the mapping of one synthetic density onto another. Figure 1 shows a mass distribution μ_0 on Ω_0, with dark regions representing little mass, lighter regions representing more. Similarly, Figure 2 indicates the density μ_1 on Ω_1. Figure 3 represents the initial mapping u, which was obtained by the method described in Section 3. The shading in this figure represents the Jacobian of u. Figure 4 shows the optimal Monge–Kantorovich mapping $\tilde{u}_{MK}$, obtained using the non-local first order equation (22). One can see that the effect of removing the curl is to straighten out the grid lines some what. Also, the shading indicates that the MP condition (1) is indeed working. On a Sparc Ultra10, this process took just a few seconds.

We also tested the method by optimally mapping μ_0, a 256 x 256 "Lena" image, onto μ_1, a "Tiffany" image. In Figure 5, we have the original Lena image. It has been histogrammed, and its values have been scaled from 0.25 to 4.0. In Figure 6, we show the Tiffany image histogrammed and scaled in the same manner as Lena. The initial mapping u is shown in Figure 7. The shading in this example differs from the previous one in that we have simply taken the Lena pixel values μ_0 and transported them via the mapping u. Here, the Jacobian of u ranges between 1/16 and 16. Next, in Figure 8, we show the Monge–Kantorovich optimal mapping $\tilde{u}_{MK}$, again found by applying the non-local flow (22). The Monge–Kantorovich mapping can be seen to be much more regular, and indicates the potential of the method. The processing of 1000 iterations took 20 minutes.

Finally, we use the warp function (26) to find a continuous transformation of Lena to Tiffany. This is illustrated in Figures 9 through 13, which correspond to $t = 0.2,\ 0.4,\ 0.6,\ 0.8$ and 1.0. Here, we have shaded the target points by $|DX^{-1}|\ \mu_0 \circ X^{-1}$, which can be seen by (1) to vary smoothly between Lena at time 0 and Tiffany at time 1.

We have also successfully implemented the second order local flow (23), with similar results. In this case, we require that a periodic boundary condition be enforced, specifically that $\tilde{u} - x$ be periodic on the square image domain. We also used an upwinding scheme when calculating $D\tilde{u}$. While it may seem that this local flow should provide a faster method than the non-local flow (22), in practice this does not seem to be the case. Even though the non-local method requires that the Laplacian be inverted during each iteration, the problem has been set up to allow the use of fast numerical solvers which use FFT-type methods and operate on rectangular grids [20]. We have used the Matlab solver here, which uses sine transforms followed by the solution of a tri-diagonal system. Moreover, we have found that the functional is decreased substantially more during each iteration of the non-local method, using the maximum temporal step size allowed for stability in each case.

In general, the target domain Ω_1 need not be rectangular when using the non-local method. However, we note that if the periodic boundary condition described above is used, then the Laplacian in (22) can be inverted using the FFT alone, without the need to solve a subsequent matrix system. For the Lena to Tiffany warp, this reduced the processing time by 1/3.

Acknowledgements. The authors would like to thank Professor Rob McCann of the University of Toronto, Professor Mike Cullen of the ECMWF, and Professors Wilfrid Gangbo and Anthony Yezzi of Georgia Tech for some very interesting conversations about the Monge–Kantorovich problem.

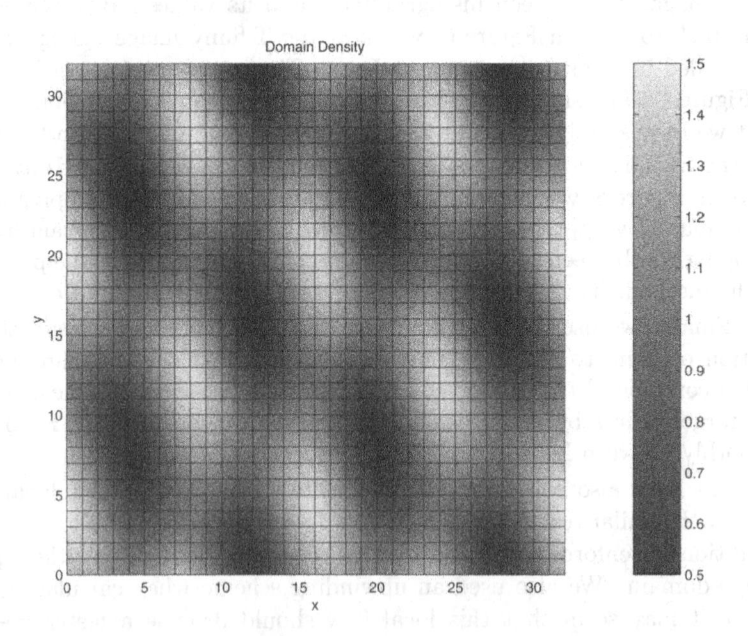

FIG. 1. *Density μ_0 on Ω_0.*

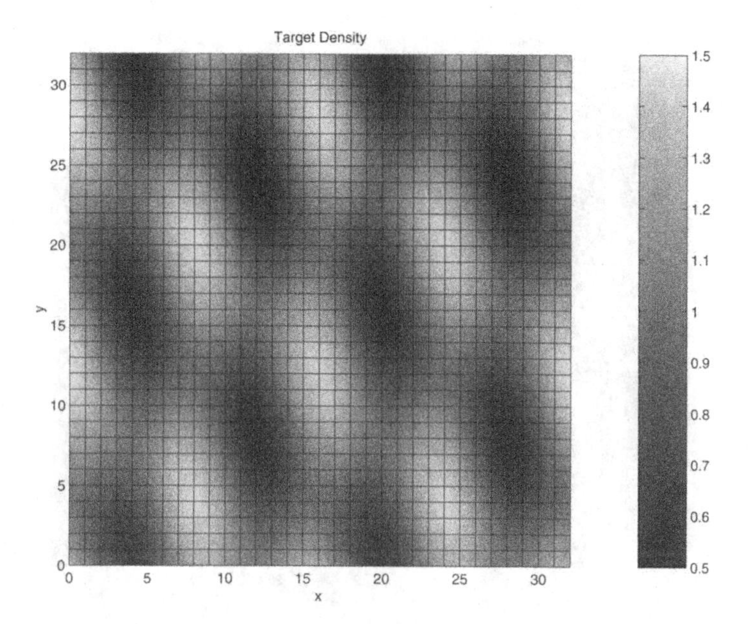

FIG. 2. *Density* μ_1 *on* Ω_1.

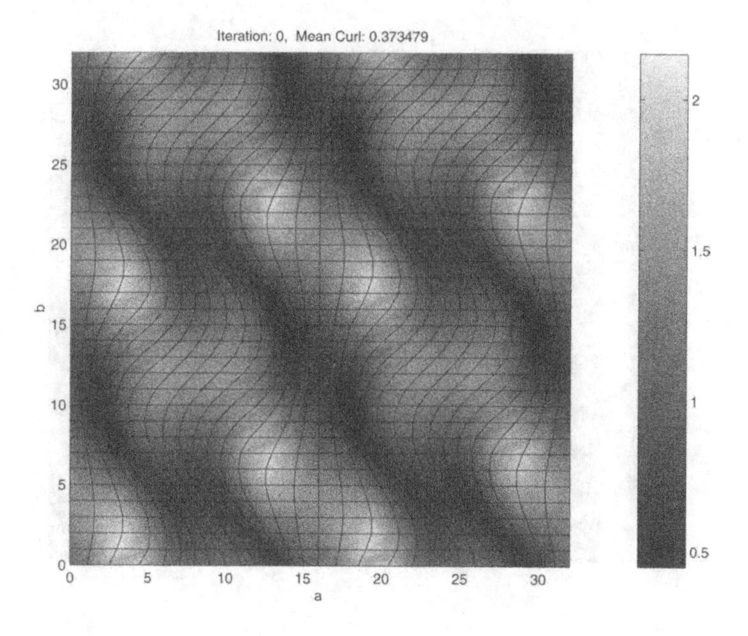

FIG. 3. *Initial mapping from* Ω_0 *to* Ω_1.

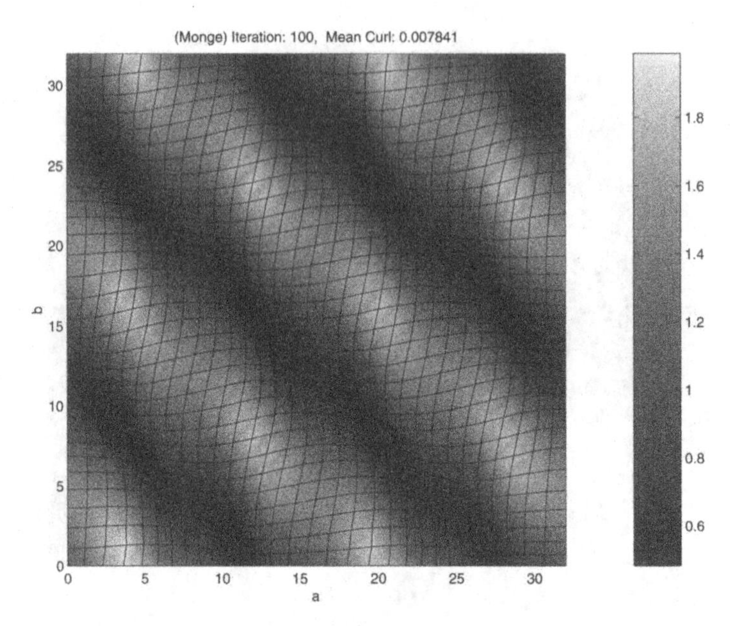

FIG. 4. *Final Monge–Kantorovich mapping from Ω_0 to Ω_1.*

FIG. 5. *Original Lena image.*

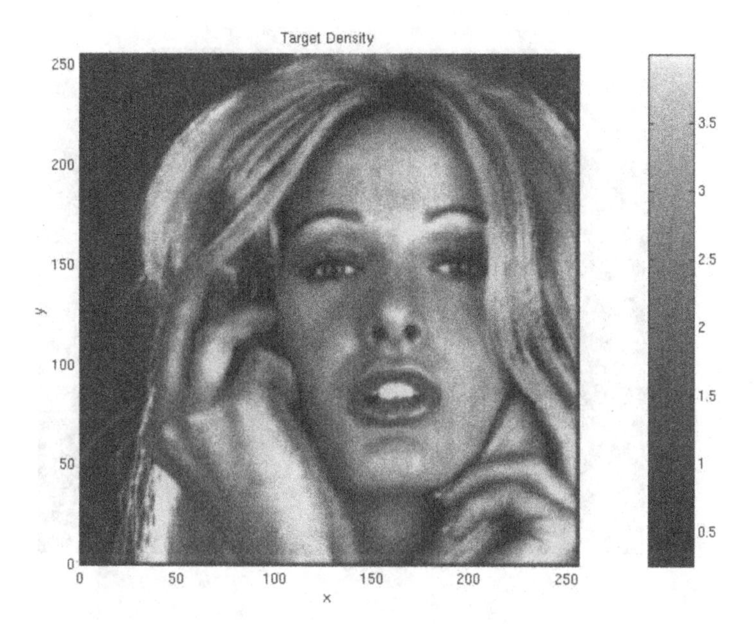

FIG. 6. *Target Tiffany image.*

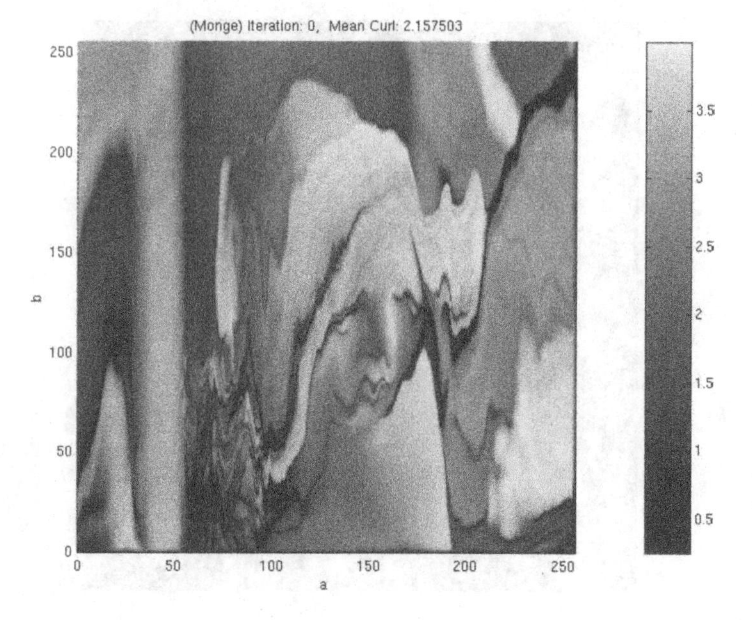

FIG. 7. *Initial mapping.*

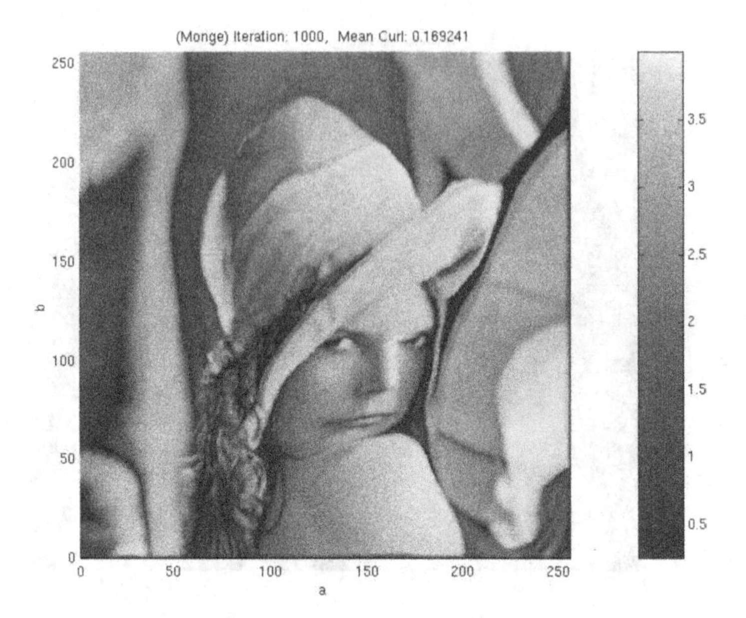

FIG. 8. *Final Monge–Kantorovich mapping.*

FIG. 9. *Lena to Tiffany warp:* $t = 0.2$.

FIG. 10. *Lena to Tiffany warp:* $t = 0.4$.

FIG. 11. *Lena to Tiffany warp:* $t = 0.6$.

FIG. 12. *Lena to Tiffany warp:* $t = 0.8$.

FIG. 13. *Lena to Tiffany warp:* $t = 1.0$.

REFERENCES

[1] L. AMBROSIO, "Lecture notes on optimal transport theory," *CIME Series of Springer Lecture Notes*, lectures given in Madeira (PT) at the Euro Summer School "Mathematical Aspects of Evolving Interfaces," July 2000.

[2] S. ANGENENT, S. HAKER, AND A. TANNENBAUM, "Minimizing flows for the Monge–Kantorovich problem," in preparation.

[3] S. ANGENENT, S. HAKER, A. TANNENBAUM, AND R. KIKINIS, "On area preserving maps of minimal distortion," in *System Theory: Modeling, Analysis, and Control*, edited by T. Djaferis and I. Schick, Kluwer, Holland, 1999, pp. 275–287.

[4] S. ANGENENT, S. HAKER, A. TANNENBAUM, AND R. KIKINIS, "Laplace-Beltrami operator and brain surface flattening," *IEEE Trans. on Medical Imaging* **18** (1999), pp. 700–711.

[5] J.-D. BENAMOU AND Y. BRENIER, "A computational fluid mechanics solution to the Monge–Kantorovich mass transfer problem," *Numerische Mathematik* **84** (2000), pp. 375–393.

[6] Y. BRENIER, "Polar factorization and monotone rearrangement of vector-valued functions," *Com. Pure Appl. Math.* **64** (1991), pp. 375–417.

[7] M. CULLEN AND R. PURSER, "An extended Lagrangian theory of semigeostrophic frontogenesis," *J. Atmos. Sci.* **41**, pp. 1477–1497.

[8] B. DACOROGNA AND J. MOSER, "On a partial differential equation involving the Jacobian determinant," *Ann. Inst. H. Poincaré Anal. Non Linéaire* **7** (1990), pp. 1–26.

[9] D. FRY, *Shape Recognition Using Metrics on the Space of Shapes*, Ph.D. Thesis, Harvard University, 1993.

[10] W. GANGBO, "An elementary proof of the polar factorization of vector-valued functions," *Arch. Rational Mechanics Anal.* **128** (1994), pp. 381–399.

[11] W. GANGBO AND R. MCCANN, "The geometry of optimal transportation," *Acta Math.* **177** (1996), pp. 113–161.

[12] W. GANGBO AND R. MCCANN, "Shape recognition via Wasserstein distance," Technical Report, School of Mathematics, Georgia Institute of Technology, 1999.

[13] S. HAKER, S. ANGENENT, A. TANNENBAUM, AND R. KIKINIS, "Nondistorting flattening maps and the 3D visualization of colon CT images," *IEEE Trans. of Medical Imaging*, July 2000.

[14] R. HARALICK AND L. SHAPIRO, *Computer and Robot Vision*, Addison-Wesley, New York, 1992.

[15] L.V. KANTOROVICH, "On a problem of Monge," *Uspekhi Mat. Nauk.* **3** (1948), pp. 225–226.

[16] M. KNOTT AND C. SMITH, "On the optimal mapping of distributions," *J. Optim. Theory* **43** (1984), pp. 39–49.

[17] R. MCCANN, "Polar factorization of maps on Riemannian manifolds," preprint 2000. Available on http://www.math.toronto.edu/mccann.

[18] R. MCCANN, "A convexity principle for interacting gases," *Adv. Math.* **128** (1997), pp. 153–179.

[19] J. MOSER, "On the volume elements on a manifold," *Trans. Amer. Math. Soc.* **120** (1965), pp. 286–294.

[20] W. PRESS, S. TEUKOLSKY, W. VETTERLING, AND B. FLANNERY, *Numerical Recipes in C: The Art of Scientific Computing*, 2nd Edition, Cambridge University Press, Cambridge U.K., 1992.

[21] S. RACHEV AND L. RÜSCHENDORF, *Mass Transportation Problems*, Volumes I and II, Probability and Its Applications, Springer, New York, 1998.

[22] G. STRANG, *Introduction to Applied Mathematics*, Wellesley-Cambridge Press, Wellesley, Mass., 1986.

[23] A. TOGA, *Brain Warping*, Academic Press, San Diego, 1999.

[24] P. VIOLA AND S. WELLS, "Alignment by maximization of mutual information," *Int. J. Computer Vision* **24** (1997), pp. 137–154.

ANALYSIS AND SYNTHESIS OF VISUAL IMAGES IN THE BRAIN: EVIDENCE FOR PATTERN THEORY*

TAI SING LEE†

Abstract. At each moment in time, we perceive a very small fragment of the world through our retinas, yet our subjective perception of the visual world in front of us is rather clear, coherent and complete. Often we see things that are not even there. This is because what we perceive is actually a 'virtual' visual world that is created in our minds – a product of the interaction between our experience, prior knowledge and the incoming sensory data. This world is dynamic and plastic. It depends on the behavioral demands imposed on us and the statistics of our experiences. In this lecture, I will present neurophysiological evidence that suggests that the early visual cortex participates in many levels of visual processing underlying the generation and the representation of this subjective visual world in our brain.

Key words. Vision, neurobiology, computational vision.

AMS(MOS) subject classifications. 68T45, 92C20.

1. Theory.

1.1. The nature of perception. The visual world we perceive is a mental construction inside our brain, rather than the raw spots and dots that photons create on our retinas. This mental construction is so real and compelling that we rarely question or think twice about it. We realize this fact through the painful examples of patients who have 'visual form agnosia'. These patients lost their ability to organize and construct objects in the virtual visual world in their minds because of lesions in their visual areas. For example, Benson and Greenberg (1969) reported a patient whose vision is normal in discriminating fine features, color and motion but couldn't put all the tiny details back together to experience and perceive coherent objects.

The raw image sampled by the retina at each of our glances provides a very impoverished image of the outside world. Figure 1 illustrates a sequence of images approximating what one of our retinas see at several fixations. These images are severely limited – they are high-resolution only in the fovea but are very blurry in the periphery. Yet, we do not 'feel' the fuzziness in the surround, we realize it only when we pay attention to it and ponder about it. However, by making saccadic eye movements three or four times a second to constantly scan the visual scene, we somehow are able to obtain enough samples of the external world to create an apparently stable and complete visual world inside our brain.

*The work was supported in part by NSF CAREER 9720350 and NIH EY08098.

†Center for the Neural Basis of Cognition and Department of Computer Science, Carnegie Mellon University, Pittsburgh, PA 15213.

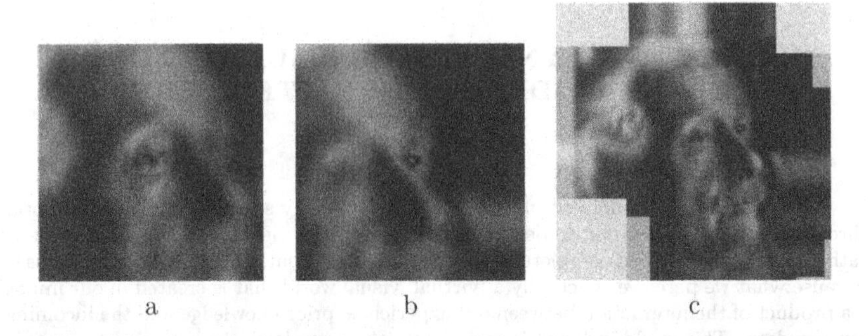

FIG. 1. *(a) and (b) Raw input sampled by the retina in two fixations. (c) A 'mental' image created in our perception by integrating the retinal images from several fixations (see also Lee and Yu 2000).*

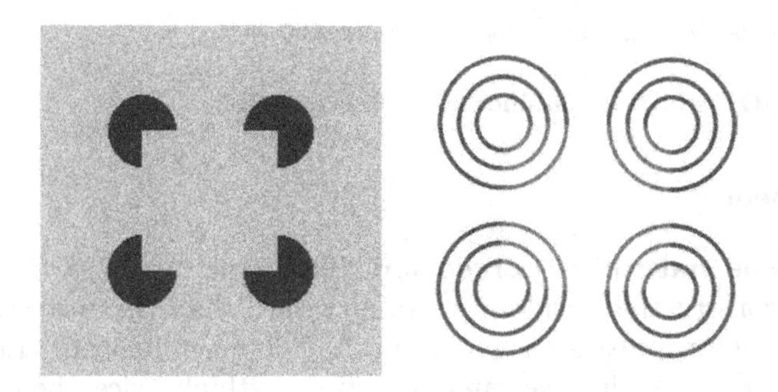

FIG. 2. *Illusory squares: we see subjective contour and surface at places where there are actually no direct visual evidence for it.*

Another piece of evidence supporting the idea of constructive processes in vision is beautifully illustrated by Kanizsa with his famous visual illusion. When viewing the display shown in Fig. 2a, we perceive a subjective square and we see vivid borders of the square even in regions of the image where there is no direct visual evidence for them. This is one example of the phenomenon of illusory figure. Figure 2b show an even more stunning example presented by Hoffman (1998). In this display, when parts of the rings change their color from black to blue, a subjective perception of a ghostly blue square surface is induced over the empty space.

1.2. Generative processes in unconscious inference. Helmholtz (1867) had argued that perception is a product of *unconscious inference*: what we perceive is our visual system's best guess as to what is in the world. This guess is based both on our prior experience and the retinal image. Can this unconscious inference be accomplished simply through association and memory? Or does it require the generation of an explicit representation in our brain?

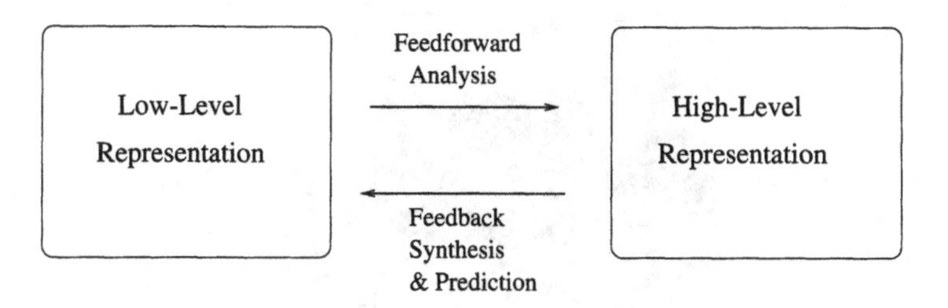

FIG. 3. *According to Grenander, inference is made through a combination of analysis and synthesis loops.*

Marr (1981) would have favored an explicit representation, but he would also say that it can be computed by a feedfoward chain of computational modules, each computing a more complex and abstract perceptual structure. The perceived abstract structure, such as the illusory surface and contour of the square, could be computed and represented by higher visual areas. There is no need to reconstruct and represent them explicitly in the early visual cortex.

On the other hand, Grenander (1976-81) would have argued that inference is accomplished through the interaction of analysis and synthesis. From this point of view, as articulated by Mumford (1992) and Lee et al. (1998), vision is a series of interactive hypothesis testing. Prediction and expectation continuously generated by the higher visual areas are tested and matched with the representations in the earlier visual areas (Figure 3). This feedback synthesis serves two purposes. First, it is useful for analyzing ambiguous images in which a dialog between knowledge and perception is required to disambiguate the scene, as in the example devised by R.C. James in Figure 4. Second, having top-down expectation and prediction is important for speeding up the inference process in real time. That is, if we know what we are going to see, it is much easier and more efficient to verify objects in a visual scene than to deduce them from sketch at each moment in time. Vision is then considered an active process of generating and testing hypotheses, very much like conducting a scientific experiment. This construction is *unconscious* and the hypotheses constructed are our perception of the visual world. Current important theories on brain functions such as Grossberg's adaptive resonance theory (1987), and McClelland and Rumelhart's (1981) interactive activation theory, Dayan et al.'s Helmholtz machine (1995), and Rao and Ballard's (2001) predictive coding model basically advocated the same fundamental view, particularly for the purpose of disambiguation.

1.3. High-resolution buffer hypothesis. Mumford and I (Lee and Mumford 1996, Lee et al. 1998) have suggested a new framework for con-

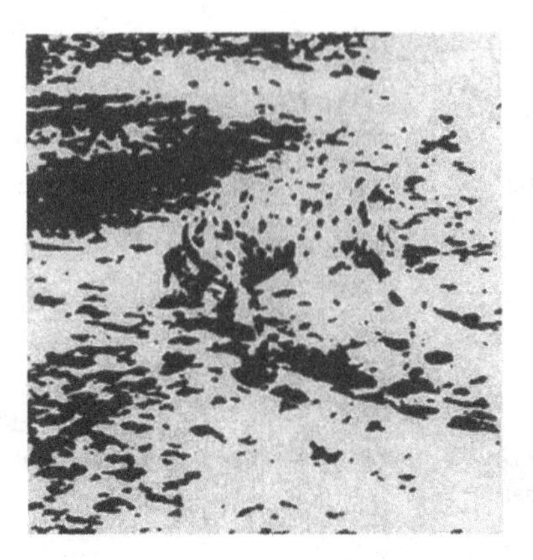

FIG. 4. *An image devised by R.C. James to illustrate how the interpretation of some images relies on top-down knowledge.*

ceptualizing the role of the primary visual cortex (V1) in visual processing from this perspective. This very large region in the occipital cortex has been traditionally considered the first stage of visual processing, extracting edges and other low-level cues. We think that it might play a far more important role than previously imagined. Because the receptive fields of neurons in V1 are much smaller and more spatially localized than those of neurons in the extrastriate cortex, V1 could furnish a unique *high resolution buffer* or a sketch pad for the whole visual cortex to make detailed geometric calculations and synthesize images through the generative processes. For example, suppose we want to explicitly construct the precise contour of the illusory square (Figure 2), V1 is an ideal place to do so because it furnishes a precise representation for integrating the bottom-up information from the raw images and the top-down hypotheses, generated from prior experience, to construct and represent the sharp subjective contour. As another example, suppose the brain needs to compute the axis of symmetry of an object for shape discrimination; V1 could provide an appropriate buffer for representing the axis explicitly. As in the case of illusory contour, the axis of symmetry is a computation that requires the integration of local information and global scene context.

More generally, we think that V1 is not limited to curvilinear geometric computation as illustrated by the examples of illusory contour and symmetry axes. Rather it serves as a high-resolution buffer for even more general computation. We know that the information output by V1 is channeled into the dorsal stream (or commonly known as the *where* pathway) for motion processing and spatial analysis, as well as the ventral stream (or commonly known as the *what* pathway) for form processing and object analysis. These two streams are further subdivided into multiple modules or areas, each responsible for processing different aspects of the visual scene: color, form, motion, stereo, and spatial locations of objects in various coordinates. A major question is how the brain combines all the processed information back together to form an unified percept. There are at least three possible loci of interaction for such an unification to occur. First, they could be mediated by the intercortical connections between modules in the two streams (Baizer et al. 1991). Second, they could be mediated in the prefrontal cortex such as area 46 where both the dorsal and ventral streams converge to (Rao et al. 1997). Third, with the massive feedforward and feedback connections it has with many extrastriate areas, V1 can potentially serve as a sketch pad for integrating the higher level information, derived from the different modules, including color, shape, depth, object identity and spatial location. The high-resolution buffer hypothesis basically argues for the importance of this third possible locus of interaction and emphasizes that V1 participates in all levels of perceptual computations that require high resolution image details and spatial precision.

2. Experiments. What evidence supports the high-resolution buffer hypothesis and, more generally, the generative and constructive processes in the brain? I will describe three experimental observations we made that are in part both supportive and suggestive of these ideas.

2.1. Illusory contour. The first experiment was to examine whether V1 represents subjective contours of the Kanizsa square as shown in Figure 2. Seventeen years ago, von der Heydt and his colleagues (von der Heydt et al, 1984) found that neurons in macaque V2 are sensitive to an illusory bar moving across their receptive fields. This discovery was seminal because it showed neurons possess a direct physiological correlate of a perceptual phenomenon. Curiously, they didn't found V1 neurons responding to the illusory contour. Hence, they proposed a feedforward model that integrates end-stopping signals in V1 to produce the illusory contour responses in V2. In short, their evidence caused a problem for all the interactive models of visual processing, as well as the high resolution buffer hypothesis which predicts that we should be able to observe the emergence of illusory contour at the later part of V1's responses because of the generative feedback processes in the visual system.

We decided to put the high-resolution buffer hypothesis to test. We (Lee and Nguyen 2001) studied the responses of V1 and V2 neurons to five sets of of stimuli, as shown in Figure 5. The set of test stimuli included a Kanizsa figure with illusory contours (Figure 5a), an amodal figure in which the contours are partially occluded (Figure 5b), and a variety of control and comparison stimuli (Figures 5c-i). In each trial, while the monkey fixated, a sequence of 4 stimuli was presented. The presentation of each stimulus in the sequence lasted for 400 msec. Over successive trials, one of the contours (real, amodal or illusory) in the figure was placed at 10 different locations relative to the center of the receptive field, 0.25^o apart, spanning a range of 2.25^o, as shown in Figure 6. It is important to bear in mind

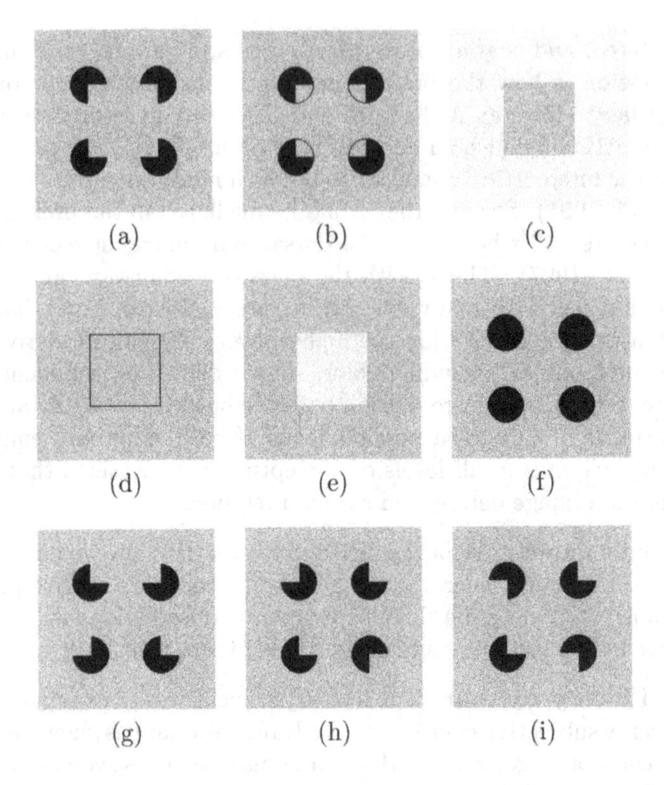

FIG. 5. *A subset of the stimuli used in the illusory contour experiment by Lee and Nguyen (2001).*

that the receptive fields of the neurons, as plotted by small oriented bars, was less than 1 degree at that eccentricity (about 2 -3 degree away from the fovea). The gap between the pac-men was 2 degree wide. The neurons are considered to be sensitive to illusory contour if their response to the illusory contour, at the precise location of that contour, was significantly larger than their response to the amodal contour or other controls. We found that a significant number of V1 neurons at the superficial layer of V1 exhibited sensitivity to the illusory contour under our experimental manipulation (Lee and Nguyen 2001).

Figure 7 presents the findings from a V1 neuron. This cell responded significantly more to the illusory contour than to the amodal condition, or the rotated corner disc configuration. The illusory contour elicited a response precisely at the same location at which a real contour elicited the maximum response. However, the response to the illusory contour appeared at 100 msec after the onset of the subjective square, as compared to 45 msec after the appearance of the square defined by lines or luminance contrast. The averaged temporal response of 50 neurons in the superficial layers of V1 to the illusory contour and to the controls (Figure 8) demon-

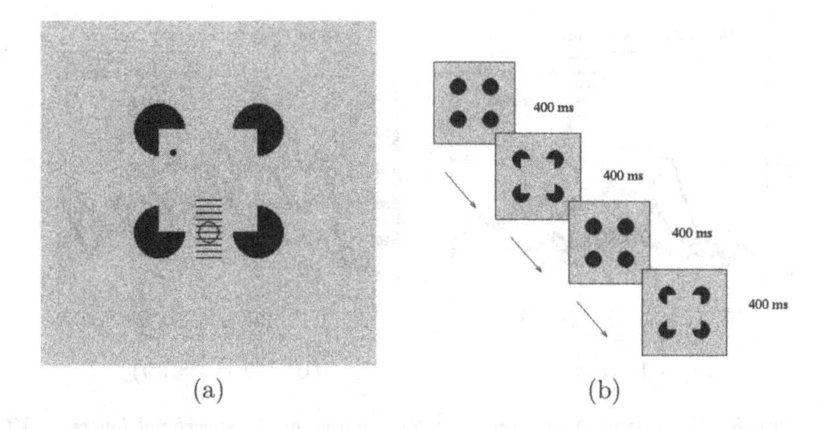

FIG. 6. *(a) The 10 different positions where the receptive field of a cell was placed relative to the subjective contour. (b) Sequence of presentation: Abrupt onset of the subjective square in front of the four discs helps to call attention to the square (Lee and Nguyen, 2001).*

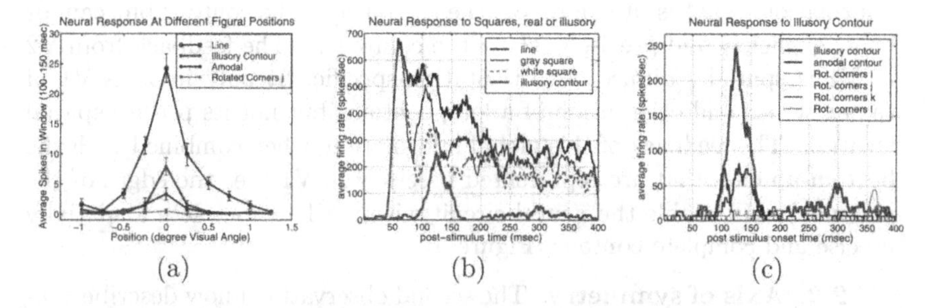

FIG. 7. *Response of a V1 neuron. (a) Response at different spatial location relative to the illusory contour. (b) The onset of response to illusory contour emerges at 100 msec, 60 msec later than the response to the real contours. (c) Response to the illusory contour (Figure 5a) is significantly greater than the response to the amodal contour (Figure 5b) and the other rotated pac-men controls (Figure 5g–i) (Lee and Nguyen, 2001).*

strated a significant response to the illusory contours at the population level. The average temporal response of 39 neurons in the superficial layers of V2 showed an earlier onset of illusory contour response. We will refer the readers to our paper (Lee and Nguyen 2001) for the details of the experiment. Suffice to say, the experimental finding suggests the illusory contour is being explicitly 'constructed' and represented in V1.

If V2 neurons are already detecting and encoding information about illusory contours, what is the advantage of feeding it back to V1? One reason is that V2 neurons' receptive fields are twice the diameter of V1 neurons at the same spatial location. Hence, V2 neurons can integrate information more globally, but it can no longer represent precise spatial

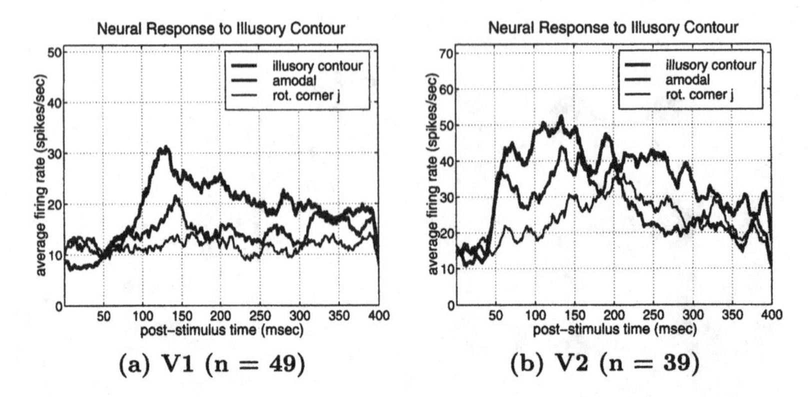

FIG. 8. *Comparing the responses of the neurons at the superficial layers of V1 and V2 to the stimuli, we can see the illusory contour signal, which is indicated by the difference between the response to illusory contour over the response to amodal or rotated pac-men control, emerges 100 msec after stimulus onset for V1 neurons, and at 65 msec for V2 neurons (Lee and Nguyen, 2001, with permission from PNAS).*

information. That is, it can detect the existence of the contour but cannot know explicitly and precisely where the contour is. The feedback from V2 to V1 is spatially diffuse but orientation-specific. It only informs V1 of the existence and orientation of a long contour, but not its precise spatial location. The feedback of this global information, when combined with the bottom-up cues that are represented precisely in V1 (i.e. the edges of the pac-men), will enable the neural circuit within V1 to complete a spatially precise and complete contour (Figure 9).

2.2. Axis of symmetry. The second observation I now describe concerns evidence suggestive of a possible medial axis representation at V1. Medial axis transform, or skeletonization, is a powerful way of representing shape. Blum (1973) proposed to describe the complex biological forms using the skeleton and a small finite set of shape primitives. The skeleton links these elementary parts together hierarchically like the clauses in a sentence. This method is particularly useful for encoding the infinite variety of biological forms, in which relationships between body parts possessing flexible joints can change drastically along with changes in view point and motion. Blum suggested a region-based description using skeletons or axis of symmetry of the objects might be more robust and stable than a boundary based description against such variations (see Figure 14).

Our experiment (Lee et al. 1998) was designed to investigate the neural representation of texture contours and surface. We examined the responses of the neurons to the strip stimuli defined by texture contrast (Figure 10). The texture defined strip was presented in a randomized series of sampling positions relative to the cell's receptive field. The cell's response at one position was sampled at each trial. In each recording session, the cell's response was sampled along a horizontal cross-section of the image at a

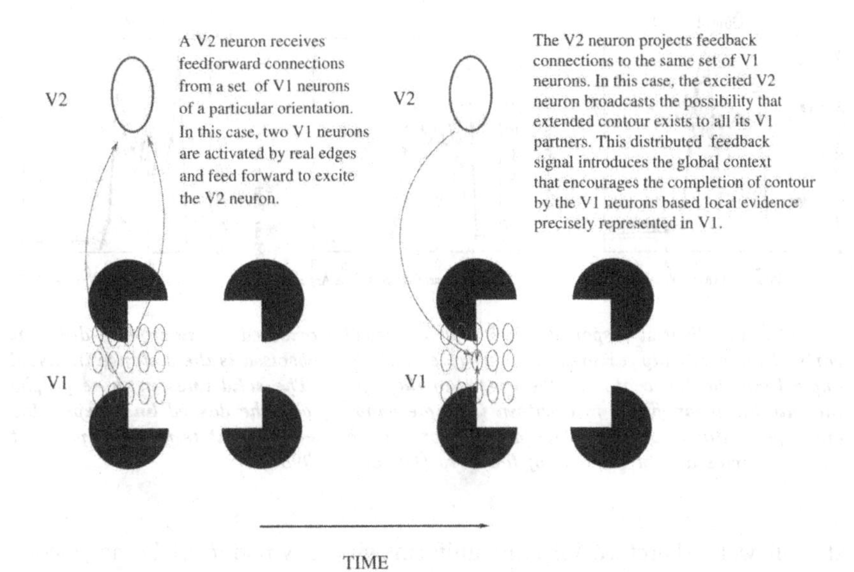

V2
A V2 neuron receives feedforward connections from a set of V1 neurons of a particular orientation. In this case, two V1 neurons are activated by real edges and feed forward to excite the V2 neuron.
V1
V2
The V2 neuron projects feedback connections to the same set of V1 neurons. In this case, the excited V2 neuron broadcasts the possibility that extended contour exists to all its V1 partners. This distributed feedback signal introduces the global context that encourages the completion of contour by the V1 neurons based local evidence precisely represented in V1.
V1
TIME

FIG. 9. *How V2 facilitates precise contour computation in V1.*

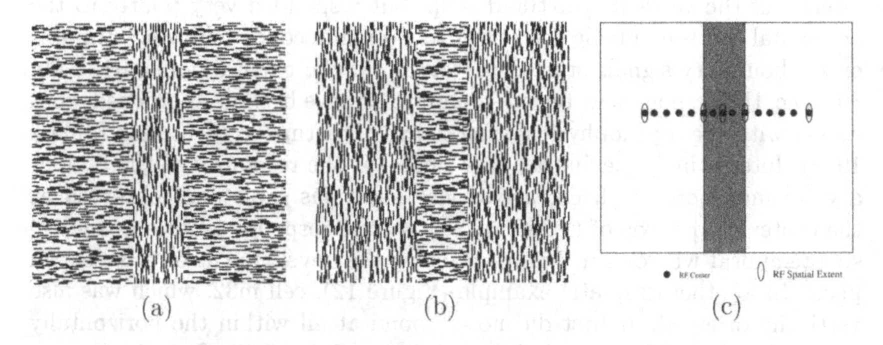

FIG. 10. *(a) A vertically textured strip in front of a horizontally textured background. (b) A 4 degree width horizontally textured strip in front of a vertically textured background. Both strip widths were 4 degree visual angle, which is 4–6 times larger than the diameter of the classical receptive fields of the cells.(c) The receptive fields of the neurons were placed at 16 locations in each of the stimulus images.*

0.5^o visual angle step interval (Figure 10c). The size of the receptive field of the cell ranged from 0.5^o to 0.8^o in visual angle.

We found that there were several stages in the responses of V1 neurons, each with distinct spatial response profiles at different temporal windows. Typically, V1 neurons started to respond about 40 msec after the stimulus was displayed on the screen. From 40 to 60 milliseconds after stimulus onset, the cells behaved essentially as local feature detectors or linear filters (Hubel and Wiesel 1978, Pollen et al 1989). The responses to the texture

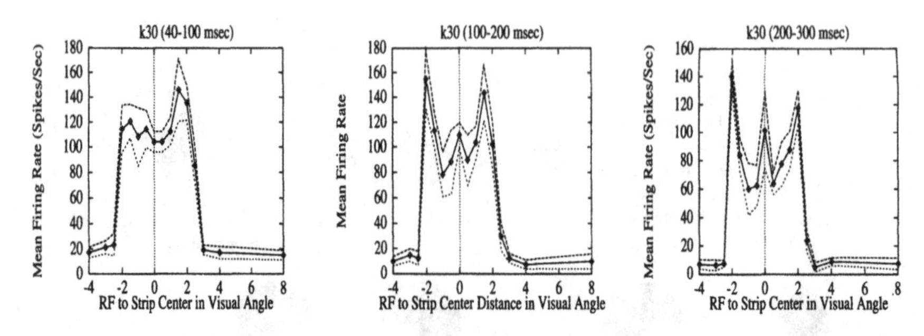

FIG. 11. *Spatial response profiles of a vertically oriented V1 neuron to different parts of the vertically textured strip (Figure 10a). The abscissa is the distance in visual angle from the RF center to the center of the figure. The solid lines in these graphs indicate the mean firing rate within the time window, and the dashed lines depict the envelope of standard error. The later response of the neuron exhibits response peaks at the boundaries and at the axis of the strip (Lee et al. 1998).*

stimuli were therefore initially uniform within a region of homogeneous texture, based upon the orientation tuning of the cells. In the example shown in Figure 11, the neuron showed preference for features of vertical orientation. In the initial period, it responded uniformly well within the interior of the vertically-textured strip, but responded very poorly to the horizontal texture outside the strip. At 60 milliseconds following stimulus onset, boundary signals started to develop at the texture boundaries. By 80 msec, the responses at texture boundaries have become sharpened, consistent with the psychophysical time course of texture segmentation (Julesz 1975). Interestingly, beginning at 80 msec, as the responses at the boundary became more localized, a response peak was sometimes observed at the center or the axis of the strip. The spatial response profiles in successive temporal windows in Figure 11 show the development of this central peak. In another dramatic example (Figure 12), cell m32, which was also vertically oriented, at first did not respond at all within the horizontally textured strip but became active at the axis of the strip after 80 msec.

Statistically significant central peaks were observed in 14 out of the 50 neurons tested with the vertically-textured strip and 10 cells with the horizontally-textured strip (T-test, $p < 0.05$). Figure 13 shows the average spatial temporal response of the 14 neurons that are sensitive to the positively textured strips, revealing a subtle response peak at the axis of symmetry of the strip.

Whether this signal that we accidentally observed has anything to do with the axis of symmetry representation is still open to question. There were two major problems related to this interpretation. First, the axis response seems to be dependent on the width of the strips, i.e. the axis response of a cell disappeared when the width of the strip was increased in all cases. This suggests it could be explained as a product of some global surround lateral inhibition and dis-inhibition mechanism. We have pointed

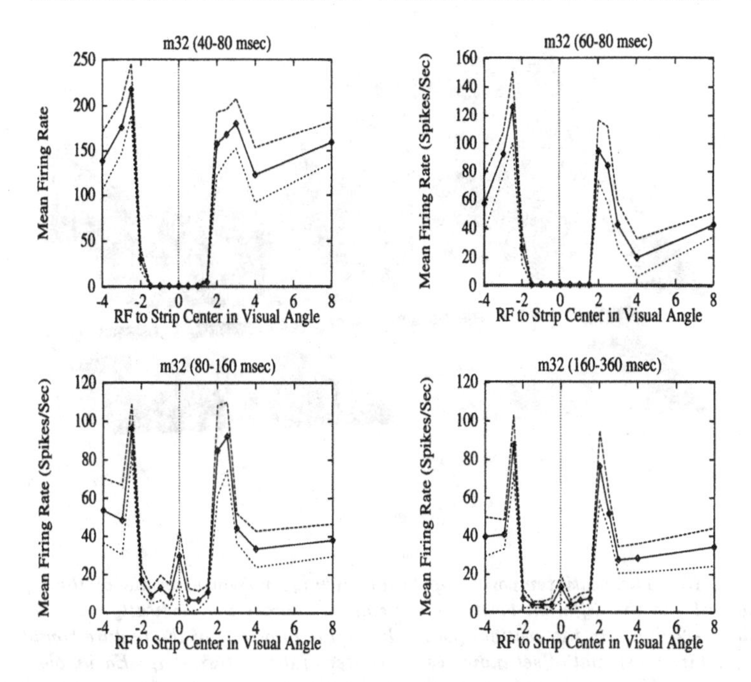

FIG. 12. *Another neuron's (cell m32) response to the horizontally textured strip (Figure 10b). Approximately 40–60 msec after stimulus onset, the cell responded uniformly to the background but did not respond to the texture strip at all because it was not tuned to the texture inside. From 60 to 80 msec, the boundary started to sharpen, but there was still no response within the strip. Interestingly, 80 msec onward, a pronounced response peak gradually developed at the axis of the texture strip (Lee et al. 1998).*

this out (Lee et al. 1998), and Zhaoping Li (1999) has also demonstrated this possibility with a model based on lateral inhibition. The fact that the neuron is sensitive to the width of the strip might not be as devastating as it might seem, for a medial axis neuron can be sensitive to the diameter of the inscribing disk as well (Figure 14). One can conceive that a group of these cells, each tuned to a particular width, could together provide an invariant representation of the medial axis. A bit more troubling is the observation that the axis response seems to be absent in black and white strips (see Lee et al. 1998 for details). We speculate that perhaps the top-down feedback is weak in this case and excitation by local features is required in order for the sub-threshold axis signal to manifest itself. However, if the signal is so weak, could it possibly serve any purpose? Could we be seeing a little too much in this response peak at the center? This experiment demonstrates that texture contour is computed and represented in V1. The evidence hints at the possibility of a neural correlate of medial axis computation, even though the data do not provide solid proof that medial axis is represented in V1 explicitly. More carefully designed experiments are required to clarify this issue.

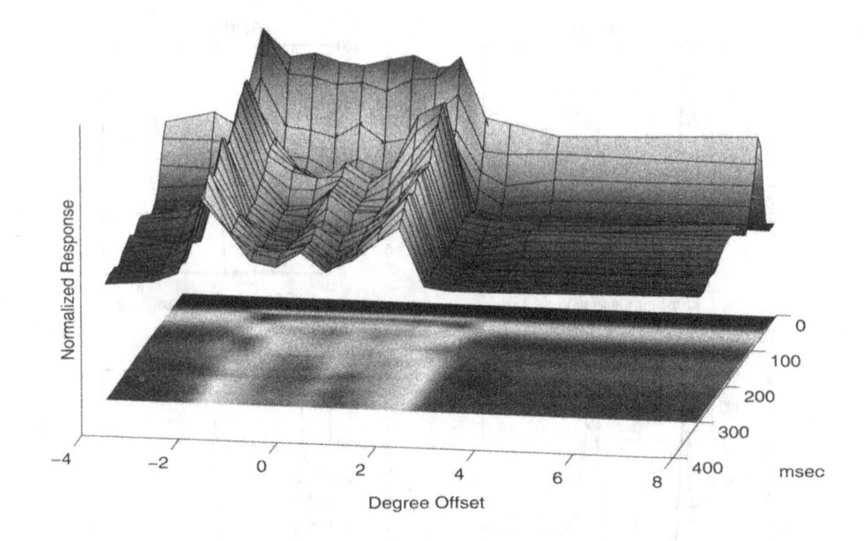

FIG. 13. *The spatiotemporal population average response profile of the 14 axis-positive cells to the different locations horizontally across the vertically textured strip. Locations −2 and 2 in the spatial offset indicate the locations of the texture boundaries. Location 0 in the spatial offset indicates the center of the texture strip. Easily observable is the strong and sharp boundary responses and an axis response of smaller magnitude in the later part of the response (see Lee et al. 1998 for details).*

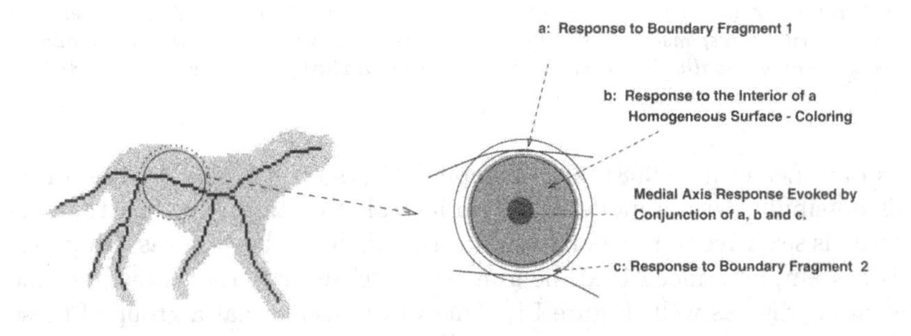

FIG. 14. *Medial axis is a descriptor that integrates local and global information. It encodes information about the location of the skeleton and the diameter of the inscribed disk. This figure illustrates how a cell may be constructed so that it fires when located on the medial axis of an object. The conjunction of three features has to be present: at least 2 distinct boundary points on a disk of a certain radius, and the homogeneity of surface qualities within an inscribing disk. Such a response is highly nonlinear, but can be robustly computed.*

2.3. Shape from shading. A third experiment, yet to be published, provides another piece of evidence in support of the generative processes. Here, we pushed the high-resolution buffer hypothesis one step further by testing whether V1 neurons are sensitive to 3D shapes. There has been

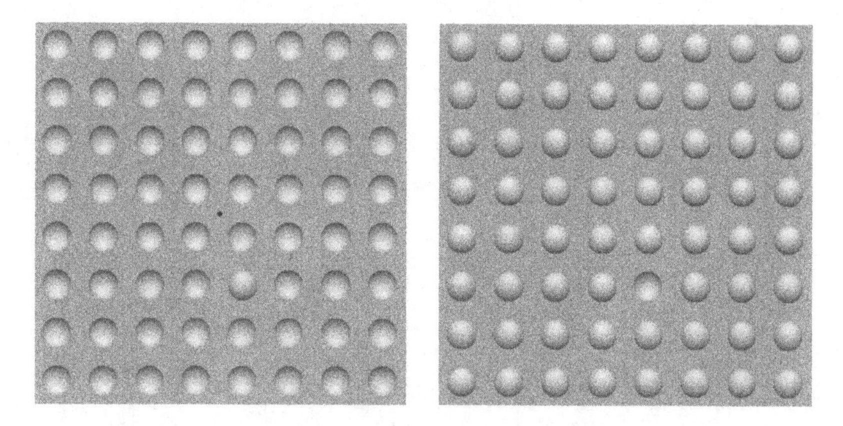

FIG. 15. *(a) a convex odd-ball stands out in a field of concavities. (b) a concave odd-ball stands out in a field of convex balls. Psychological study shows that this pop-out can be detected 'preattentively' despite the fact that the stimulus element is defined by 3D shape from shading.*

earlier physiological studies showing V1 neurons are sensitive to perceptual pop-out in terms of oriented bar or oriented texture (Knierim and Van Essen 1993, Lamme 1995, Lee et al 1998). That is, a cell responded better when its receptive field was seeing a bar(s) of optimal orientation as it was surrounded by bars of an orthogonal orientation than when it was surrounded by bars of the same orientation. But perceptual pop-out is not limited to oriented bars. Figure 15 showed examples in which a 3D convex shape pops out from a set of 3D concave shapes and vice versa (Ramachrandron 1988, Sun and Perona 1996). This experiment was to designed to examine whether V1 neurons are sensitive to odd-ball pop-out, defined by 3D shape from shading, at a relatively high level construct.

We trained the monkeys to make eye movement towards the odd-ball which could appear at one of the four random positions. The pop-out stimuli include 3D shape stimuli as well as 2D luminance contrast stimuli, in which stimulus elements were arranged in an 8×8 array (3×3 arrays are shown Figure 16 as iconic examples). Then we tested whether a V1 neuron, when its receptive field was placed inside one of the stimulus elements, is sensitive to the difference in the surrounding context; i.e., the V1 neuron responded differently when it was surrounded by dissimilar elements (pop-out condition) as opposed to when it was surrounded by similar elements (homogeneous condition).

We found that V1 neurons exhibit neural pop-out response, defined as the relative increase in response to the pop-out condition over the homogeneous condition of the shape from shading stimuli (Figure 17). The pop-out response for 3D shapes (Figure 17a,b) was significantly greater than that for 2D luminance contrast stimuli (Figure 17c,d). The pop-out

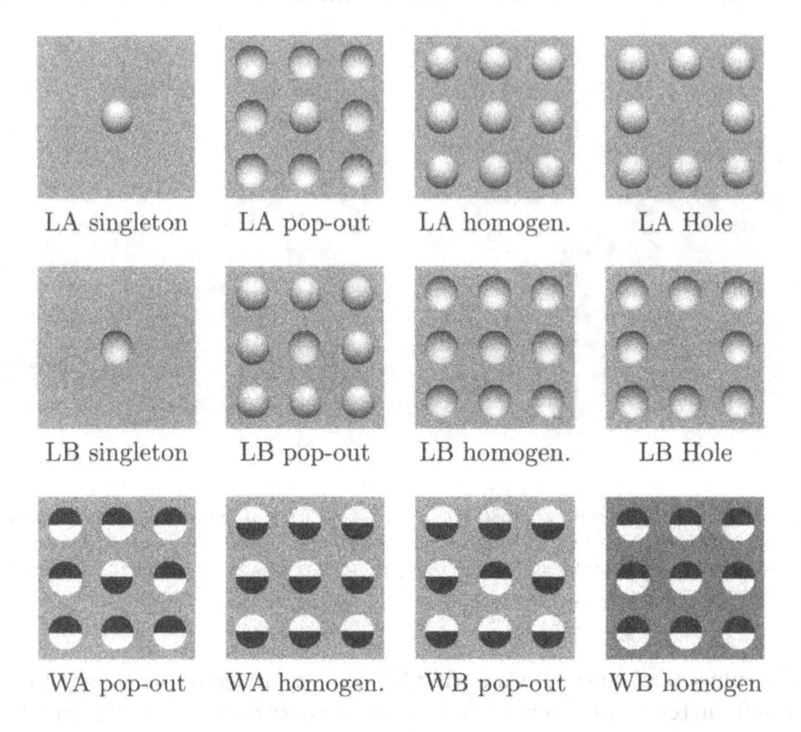

FIG. 16. *LA: lighting from above; LB: lighting from below; WA: white above; WB: white below. These are iconic representations of the stimuli. The actual stimulus display is an array of 12 × 12 stimulus elements as shown in Figure 15. In this iconic representation, the receptive field of the neuron is placed inside the stimulus at the center element (or the hole). The sizes of the classical receptive fields of the neurons (minimum responsive area) we studied are smaller than the center element.*

response is correlated with the monkey's performance in the pop-out detection test. When the pop-out signal was stronger, the monkey reacted faster and more accurately (see behavioral performance data below the figures). More interestingly, when we manipulated the statistics of the occurrence of the pop-out stimulus – specifically, when the convex pop-out was presented more frequently than any other odd-balls, we found that the behavioral preference of the monkeys shifted, and the relative neural pop-out response among the different stimuli also changed accordingly (Figure 18). This finding suggests that 1) V1's pop-out response is sensitive to 3D shape from shading information, 2) the behavioral relevance of the pop-out stimulus can determine the level of the pop-out response. From this observation and from similar observations from other monkeys, we think that when the monkey is looking for the concave object, as in this case, it is in part because the concavity was preattentively more salient than the other odd-balls. Hence the V1 neurons' activities are relatively more enhanced for the concave pop-out as well. This 'pop-out' response enhancement signal thus may be related to the so-called object attention.

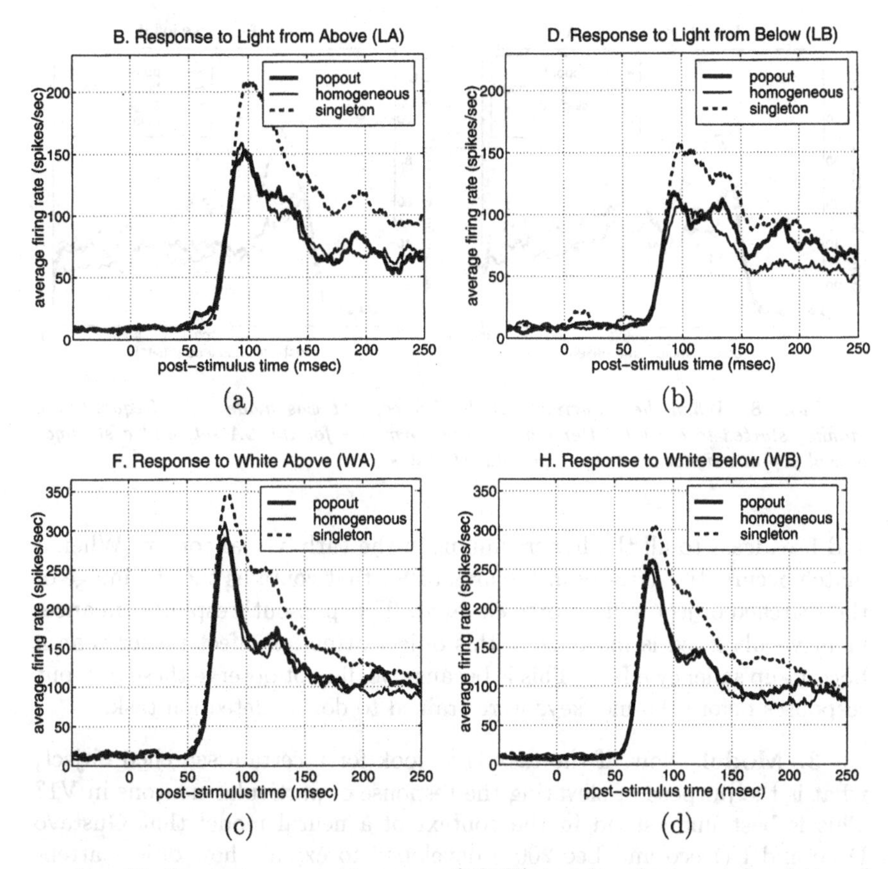

FIG. 17. *Population averaged response (PSTH) depicting the temporal evolution of the neuronal activity in V1 neurons to the four stimulus sets. Response to the pop-out is compared against the response to singleton and to a homogeneous field of convex objects. (a) convex (or lighting from above) pop-out; (b) concave (or lighting from below); (c) white above 2D luminance contrast; (d) white below 2D luminance contrast. We can observe that for this monkey, the response to the pop-out condition is stronger than the homogeneous condition in the LB set, and much less so for the LA, WA and WB sets. The behavioral performance measure associated with each pop-out condition shows that the monkey reacts faster and more accurately to the LB condition.*

Object attention is generated when an animal is searching for a particular object or feature over a large visual space in parallel (James 1890). Object attention is sometimes also called feature attention. This is in contrast with the spatial attention proposed by Helmholtz (1867) and Treisman (1982). Spatial attention can be visualized as a spotlight that 'illuminates' a certain location of visual space for focal visual analysis. When the monkeys are performing visual search for a particular object, they basically function in the object attention mode. Since object attention is object specific but spatially distributed, one can imagine that object templates are being sent down from IT (Inferotemporal cortex – an object encoding area)

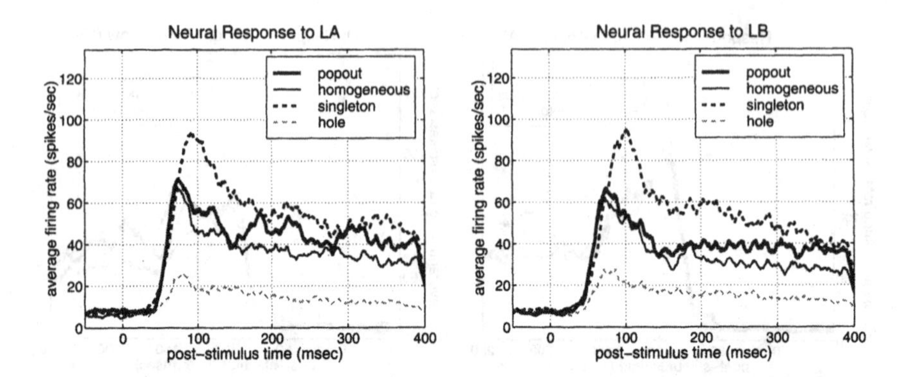

FIG. 18. *When the occurrence of the LA pop-out was made more frequent, the monkey started to exhibit better detection performance for the LA set, and a stronger neural pop-out signal for the LA pop-out stimulus.*

and broadcast to all the hypercolumns in the early visual cortex. When a match occurs, the activity in the area of V1 that contains the the image of the searched object will become elevated. The 'pop-out' response enhancement we observed is more likely this object attention effect rather than a bottom-up saliency effect. This is because we did not observe these pop-out responses before the monkeys were trained to do the detection task!

3. Model. Now, if the task is to look for a certain searched object, what is the purpose of elevating the response of particular neurons in V1? This is best understood in the context of a neural model that Gustavo Deco and I (Deco and Lee 2002) developed to explain how object attention and spatial attention would function in a unified system to accomplish translation-invariant object recognition and visual search. The premise of this model is that the early visual cortex serves as a high resolution buffer for the dorsal stream and the ventral stream to interact, combine and coordinate their information in a set of feedforward/feedback loops (Figure 19). The model is formulated in the framework of biased competition. Basically, within each cortical area, there is inhibitory competition among neurons. However, there is also excitatory facilitation between corresponding neurons across the different areas. This long range facilitation from one area can serve as a bias that will tilt the balance of competition within each module. The conceptual framework of biased competition has been proposed by Duncan and Humphrey (1989). This scheme has been implemented to explain attentional phenomenon observed in IT by Usher and Niebur (1996) and in V4 by Reynolds et al. (1999). Our contribution is to bring the dorsal stream and the early visual cortex into the picture in explaining how attentional modulation can be mediated in the ventral stream. Furthermore, our neural model also provides a functional rationale for attentional modulation. It shows attentive object recognition and vi-

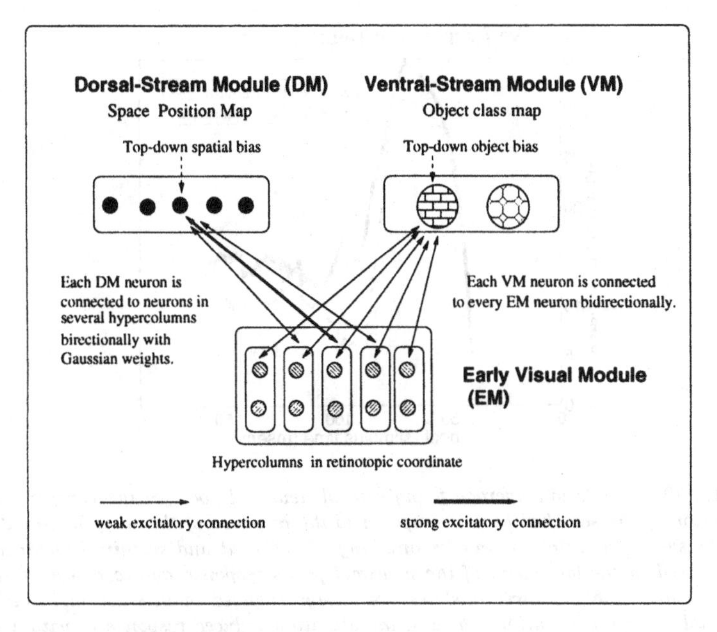

FIG. 19. *A schematic diagram of the model. The model contains three modules: the early visual module (EM), the ventral stream module (VM) and the dorsal stream module (DM). The early visual module contains orientation-selective complex cells and hypercolumns, as in the primary visual cortex. The ventral stream module contains neuronal pools encoding specific object classes, as in the inferotemporal cortex. The dorsal stream module contains a map encoding positions in the retinotopic coordinate. The early module and the ventral module are connected with symmetrical connections developed with Hebbian learning. The early module and the dorsal module are connected with symmetrically localized connections modeled with Gaussian weights. Competitive interaction within each module is mediated by inhibitory pools. Connection between modules are excitatory, providing biases for shaping the competitive dynamics within each module. Concentration of neural activities to an individual pool in the ventral module corresponds to object recognition. Concentration of neural activities to a small number of nearby pools in a dorsal module corresponds to object localization. The early module provides a buffer for the ventral and the dorsal modules to interact (Deco and Lee, 2002).*

sual search can be accomplished through the interaction of the two streams via the early visual cortex in an unified system.

The system basically has three modules: a dorsal module, which contains a spatial map for coordinating competition and coding of the location of spatial attention; a ventral module, which contains a set of neurons for coding different object classes, and an early visual module, modeled after V1. Spatial attention is initiated by introducing a top-down bias to a particular neuronal pool in the dorsal map, which will help to elevate the activities of a particular area in V1, effectively gating information from V1 to higher processing areas in the ventral stream. Alternatively, object attention – the search for a particular object is initiated by introducing

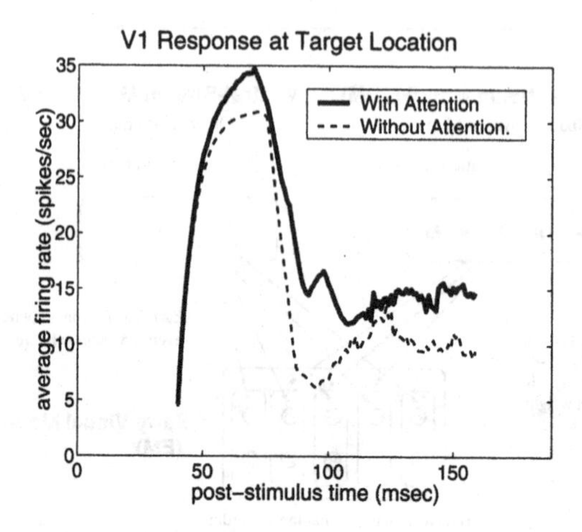

FIG. 20. *Population average firing rate of neuronal pools in the early module at the location of the searched target compared to the response of the same neuronal pools when the system was not looking for anything. Significant and sustained enhancement was observed in the later part of the neuronal pool's response due to object attention. However, given that all three modules are always engaged in interaction in either of the spatial or object attentional modes, the attention-induced response elevation in the early visual module or any other area cannot be considered as a purely spatial or a purely object attentional effect. From this point of view, all attentional effects observed necessarily possess both spatial and object attentional components.*

a top-down bias to a particular neuronal pool in the ventral module (IT). IT will then send down its expectation of V1 activities corresponding to the particular target object down to V1. A match will produce a bias in favor of that area of V1, enhancing its activities (Figure 20). The mutual coupling between V1 and the dorsal stream will allow the competition within each of these modules to help each other synergistically, thus eventually leading the contraction of activities in both V1 and the dorsal spatial map to a particular spatial location. The contraction of response to a localized area in the retinotopic space corresponds to the localization of the searched target. The reason V1 is needed in this visual search task is that in order for the monkey to make a saccadic eye movement towards the target, it has to be able to localize the object with spatial precision – provided explicitly at the level of V1. This explains why before the monkey utilized the stimuli in its behavior, we did not observe the contextual elevation in V1's activities. Only after the monkey had performed the detection task did it start to finely appreciate the stimuli both in space and in feature, leading to the sensitivity of its V1 neurons to the 3D pop-out context. This finding is again consistent with the generative processes suggested by Pattern theory.

4. Conclusion. The neurophysiological evidence presented here strongly suggests that during perception at the appropriate context explicit local and global representations might be actively constructed at the level of V1 in conjunction with the extrastriate cortices. Other neurophysiological findings, particularly the curve tracing experiment of Roelfsema et al. (1998) and a recent experiment of Paradiso and his colleagues on the effect of expectation, also point in the same direction. Visual inference is as much a generative and synthesis process as an analysis and deduction process. Our evidence suggests that low level vision and high level vision are intimately intertwined. The primary visual cortex, rather than simply contributing to the first stage of early visual processing, might play a more important role by providing a high-resolution buffer to mediate the interaction among the different expert extrastriate modules and streams. Furthermore, the primary visual cortex might actively participate in many computations that are responsible for constructing the illusion of stable and complete visual world in our mind.

REFERENCES

[1] J.S. Baizer, L.G. Ungerleider, and R. Desimone, *Organization of visual inputs to the inferior temporal and posterior parietal cortex in macaques. J. Neuroscience*, **11**(1): 168–190, 1991.

[2] D.F. Benson, D.F., and J.P. Greenberg, *Visual form agnosia: A specific defect in visual discrimination.* Archives of Neurology, **20**: 82–89, 1969.

[3] H. Blum, *Biological shape and visual science.* J. Theoretical Biology, **38**: 205–287. 1973.

[4] P. Dayan, G.E. Hinton, R.M. Neal, and R.S. Zemel, *The Helmholtz machine.* Neural Computation, **7**(5): 889–904, 1995.

[5] G. Deco and T.S. Lee, *An unified model of spatial and object attention based on inter-cortical biased competition.* Neural Computing, in Press.

[6] J. Duncan, J. and G. Humphreys, *Visual search and stimulus similarity. Psychological Review*, **96**: 433–458, 1989.

[7] U. Grenander, *Lectures in Pattern Theory I,II and III: pattern analysis, pattern synthesis and regular structures*, Springer-Verlag, 1976–1981.

[8] S. Grossberg, *Competitive learning: from interactive activation to adaptive resonance.* Cognitive Science **11**: 23–63, 1987.

[9] H.V. Helmholtz, Handbuch der physiologischen Optik. Leipzig: Voss, 1867.

[10] D.D. Hoffman, Visual intelligence: how we create what we see. W.W. Norton and Company, 1998.

[11] D.H. Hubel and T.N. Wiesel, *Functional architecture of macaque monkey visual cortex. Proc. Royal Soc. B*, (London), **198**: 1–59, 1978.

[12] B. Julesz, *Experiments in the visual perception of texture. Sci Amer.*, **232**: 34–43, 1975.

[13] J.J. Knierim, J.J. and D.C. Van Essen *Neuronal responses to static texture patterns in area V1 of the alert macaque monkey. J. Neurophysiology*, **67**: 961–980, 1992.

[14] V.A.F. Lamme, *The neurophysiology of figure-ground segregation in primary visual cortex. J. Neuroscience*, **10**: 649–669, 1995.

[15] T.S. Lee, and D. Mumford, *The role of V1 in scene segmentation and shape representation.* Society of Neuroscience Abstract, Vol. 22, 117.7, p. 283, 1996.

[16] T.S. Lee, D. Mumford, R. Romero, and V.A.F. Lamme, *The role of the primary visual cortex in higher level vision. Vision Research*, **38**(15–16): 2429–54, 1998.

[17] T.S. Lee and M. Nguyen, Dynamics of subjective contour formation in the early visual cortex. PNAS, **98**(4): 1907–1911, 2001.

[18] T.S. LEE, AND S. YU, *An information-theoretic framework for understanding saccadic eye movements.* In Advance in Neural Information Processing Systems, 12. Ed. S.A. Solla, T.K. Leen, K-R. Muller, MIT Press, 834–840, 2000.

[19] Z. LI, *Visual segmentation by contextual influences via intra-cortical interactions in the primary visual cortex.* Network **10**(2): 187–212, 1999.

[20] D. MARR, *Vision.* N.J: W.H. Freeman & Company, 1982.

[21] J.L. MCCLELLAND AND D.E. RUMELHART, *An interactive activation model of context effects in letter perception. Part I: An account of basic findings.* Psychological review: **88**: 375–407. 1981.

[22] D. MUMFORD, *On the computational architecture of the neocortex II.* Biological Cybernetics, **66**: 241–251, 1992.

[23] D.A. POLLEN, J.P. GASKA, AND L.D. JACOBSON, *Physiological constraints on models of visual cortical functions.* Models of brain function, ed. Rodney M., Cotterill, J. Cambridge University Press, 115–135, 1989.

[24] V.S. RAMACHANDRAN, *Perception of shape from shading. Nature,* **331**: 163–166 (1988).

[25] R. RAO AND D.H. BALLARD, *Predictive coding in the visual cortex: a functional interpretation of some extra-classical receptive-field effects.* Nature Neuroscience. **2**(1): 79–87, 1999.

[26] S.C. RAO, G. RAINER, AND E.K. MILLER, INTEGRATION OF WHAT AND WHERE IN THE PRIMATE PREFRONTAL CORTEX. Science, **276**(5313): 821–824.

[27] J. REYNOLDS, L. CHELAZZI, AND R. DESIMONE, *Competitive mechanisms su bserve attention in macaque areas V2 and V4. Journal of Neuroscience,* **19**: 1736–1753, 1999.

[28] P.R. ROELFSEMA, V.A. LAMME, AND H. SPEKREIJSE *Object-based attention in the primary visual cortex of the macaque monkey. Nature,* **395**(6700): 376–81. 1998.

[29] J. SUN AND P. PERONA, *Where is the sun? Nature Neuroscience,* **1**(3):183–4, 1998.

[30] A. TREISMAN AND G. GELADE, A FEATURE-INTEGRATION THEORY OF ATTENTION. *Cognitive P sychology,* **12**: 97–136 (1980).

[31] M. USHER AND E. NIEBUR, *Modelling the temporal dynamics of IT neurons in visual search: A mechanism for top-down selective attention. Journal of Cognitive Neuroscience,* **8**: 311–327, 1996.

[32] R. VON DE HEYDT, E. PETERHANS, AND G. BAUMGARTHNER. *Illusory contours and cortical neuron responses. Science* **224**(4654): 1260–1262, 1984.

NONLINEAR DIFFUSIONS AND OPTIMAL ESTIMATION*

ILYA POLLAK†

Abstract. A nonlinear diffusion process known to be effective for image segmentation and binary classification problems [12–14] is further analyzed in both 1-D and 2-D. It is shown to optimally solve certain estimation problems in 1-D and to result in an efficient ($O(N \log N)$) and exact method for solving the total variation minimization problem [16] in 1-D.

Key words. Estimation, detection, image enhancement, segmentation, total variation, nonlinear diffusions.

AMS(MOS) subject classifications. 94A08, 65K10, 62F30, 65K05, 34A12, 34A36, 65L05.

1. Introduction. The recent years have seen a great number of exciting developments in the field of nonlinear diffusion filtering of images. Many theories have been proposed that result in edge-preserving scale-spaces possessing various interesting properties. One striking feature unifying many of these frameworks is that they are deterministic. Usually, one starts with a set of "common-sense" principles which an image smoothing operation should satisfy. Examples of these are the axioms in [4] and the observation in [11] that, in order to achieve edge preservation, very little smoothing should be done at points with high gradient. From these principles, a nonlinear scale-space is derived, and then it is analyzed–again, deterministically. Note, however, that since the objective of these techniques is usually restoration or segmentation of images in the presence of noise, a natural question to ask would be:

> Do nonlinear diffusion techniques solve standard estimation or detection problems?

An affirmative answer would help us understand which technique is suited best for a particular application, and aid in designing new algorithms. It would also put the tools of the classical detection and estimation theory at our disposal for the analysis of these techniques. Attempts to address this relationship between nonlinear diffusions and optimal estimation have remained scarce–most likely, because the complex nature of the nonlinear partial differential equations (PDEs) considered and of the images of interest make this analysis prohibitively complicated. Most notable exceptions are [17, 19] which establish qualitative relations between the Perona-Malik equation [11] and gradient descent procedures for estimating random fields

*This work was supported in part by a National Science Foundation CAREER award CCR-0093105 and startup funds from Purdue Research Foundation.

†School of Electrical and Computer Engineering, Purdue University, 1285 EE Bldg., West Lafayette, IN 47907 (e-mail: ipollak@ecn.purdue.edu, WWW: http://www.ece.purdue.edu/~ipollak).

modeled by Gibbs distributions. Bayesian ideas are combined in [20] with snakes and region growing for image segmentation. In [5], concepts from robust statistics are used to modify the Perona-Malik equation.

The goal of this paper is to further explore the relationship between the nonlinear diffusion techniques and optimal estimation. We build on our contributions of [12, 14], obtaining new methods for solving certain restoration problems. The methods are first developed in 1-D, where they are provably fast, and provably optimal. While we do not have analytical results on the 2-D generalizations of our methods, experiments show that the 2-D algorithms are efficient and robust, as well.

We concentrate most of our attention on the problem of maximum likelihood (ML) estimation in additive white Gaussian noise, subject to a constraint on the total variation (TV). We show that this problem, in 1-D, is closely related to the total variation minimization problems posed by Bouman and Sauer in [6] and by Rudin, Osher, and Fatemi in [16]. This choice of our prototypical problem is motivated by a great success of total variation minimization methods [1, 2, 6, 9, 16], which has demonstrated a critical need for fast computational techniques [3, 7, 9, 18]. We will show that our 1-D method is faster (i.e., has a lower computational complexity) than the existing methods.

After a review of background material on nonlinear diffusions in Section 2, we focus on one very simple evolution in Section 3. We then show in Section 4 that this evolution results in a fast solver of our ML problem. Section 5 presents illustrations of our methods both in 1-D and 2-D.

Several proofs are only outlined and some are omitted altogether. The details of these proofs will appear in a longer paper [15], currently in preparation.

2. Background, notation, and definitions. In this section, we introduce the concepts and notation used in the remainder of the paper.

2.1. Nonlinear diffusions in 2-D. Two basic problems considered in this paper are noise removal and segmentation of noisy 1-D signals and 2-D images. By segmentation, we mean partitioning a 1-D signal or a 2-D image into homogeneous regions (these notions will be precisely defined in the course of the paper). We build on the results in [13], where a family of systems of ordinary differential equations, called Stabilized Inverse Diffusion Equations (SIDEs), was proposed for restoration, enhancement, and segmentation of signals and images. The (discretized) signal or image $\mathbf{u}^0$ to be processed is taken to be the initial condition for the equation, and the solution $\mathbf{u}(t)$ of the equation provides a fine-to-coarse family of segmentations of the image. This family is indexed by the "scale" (or "time") variable t, which assumes values from 0 to ∞. Two neighboring pixels belong to the same region at time t if their values at that time are equal to each other. Initially ($t = 0$), the finest possible segmentation is assumed: each pixel is a separate region. In the course of evolution,

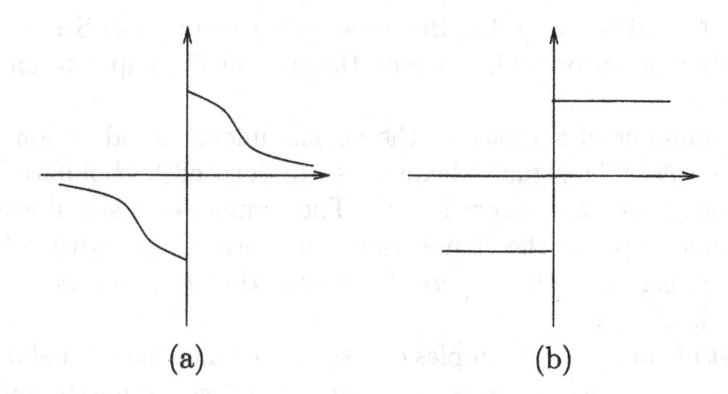

FIG. 1. *Functions F from the right-hand side of the SIDE: (a) generic form; (b) the signum function.*

two neighboring regions are merged whenever the difference between their intensity values becomes equal to zero—as shown in [13], this will occur in finite time for every pair of neighboring regions. Once two regions are merged, they stay merged for the remainder of the evolution.

We now re-state the definition of the SIDE from [13]. The intensity value u_i inside the i-th region evolves according to

$$\dot{u}_i = \frac{1}{m_i} \sum_{j \in A_i} F(u_j - u_i) p_{ij}, \tag{2.1}$$

where

$\dot{u}_i$ is the time derivative of u_i;

m_i is the number of pixels in the i-th region;

A_i is the set of the indexes of all the neighbors of the region i (in 1-D, this set consists of just one index when i is the leftmost or the rightmost region, and of two indexes otherwise);

p_{ij} is the length of the boundary between the regions i and j (always equal to one in 1-D);

F is a function which is monotonically non-increasing[1] and continuous everywhere except at zero; it is an odd function and is non-negative for positive values of its argument (see Figure 1(a)).

The usefulness of the SIDEs for image segmentation was shown in [13]; in particular, it was experimentally demonstrated that SIDEs are robust to noise outliers and blurring. They are considerably faster than other image processing algorithms based on evolution equations, since region merging reduces the dimensionality of the system during evolution. In [10], SIDEs were successfully incorporated as part of an algorithm for segmenting dermatoscopic images.

[1]In [13], we required F to be monotonically decreasing, whereas here and in [14] we slightly relax this requirement.

2.2. Notation in 1-D. Before we describe our 1-D SIDE evolution, we introduce notation which is used throughout the paper to analyze this evolution.

The number of samples in the signals under consideration is always denoted by N. The signals themselves are denoted by boldface lowercase letters and viewed as vectors in $\mathbb{R}^N$. The samples of a signal are denoted by the same letter as the signal itself, but normal face, with subscripts 1 through N, e.g., $\mathbf{u} = (u_1, \ldots, u_N)^T$. We say that u_i is the *intensity* of the i-th sample of $\mathbf{u}$.

A set of consecutive samples of a signal $\mathbf{u}$ which have equal intensities, $u_i = u_{i+1} = \ldots = u_j$, is called a *region* if this set cannot be enlarged–in other words, if

$$\begin{array}{ll} \text{either } i = 1 & \text{or } u_{i-1} \neq u_i, \text{ and} \\ \text{either } j = N & \text{or } u_j \neq u_{j+1}. \end{array}$$

Any pair of consecutive points with unequal intensities is called an *edge*. The number of distinct regions in a signal $\mathbf{u}$ is denoted by $p(\mathbf{u})$. The indexes of the left endpoints of the regions are denoted by $n_i(\mathbf{u})$, $i = 1, \ldots, p(\mathbf{u})$ (the n_i's are ordered from left to right, $n_1(\mathbf{u}) < n_2(\mathbf{u}) < \ldots < n_{p(\mathbf{u})}(\mathbf{u})$); the intensity of each sample within region i is denoted by $\bar{u}_i$, and is referred to as the *intensity* of region i. This means that $n_1(\mathbf{u})$ is always 1, and that

$$\bar{u}_i = u_{n_i(\mathbf{u})} = u_{n_i(\mathbf{u})+1} = \ldots = u_{n_{i+1}(\mathbf{u})-1}, \text{ for } i = 1, \ldots, p(\mathbf{u}),$$

where we use the convention $n_{p(\mathbf{u})+1} = N + 1$.

The length $m_i(\mathbf{u})$ of the i-th region of $\mathbf{u}$ (i.e., the number of samples in the region) satisfies:

$$m_i(\mathbf{u}) = n_{i+1}(\mathbf{u}) - n_i(\mathbf{u}).$$

Two regions are called *neighbors* if they have consecutive indexes. The number of neighbors of the i-th region of $\mathbf{u}$ is denoted by $\rho_i(\mathbf{u})$, and is equal to 1 for the leftmost and rightmost regions, and two for all other regions:

$$\rho_i(\mathbf{u}) = \begin{cases} 1, & i = 1, p(\mathbf{u}), \\ 2, & \text{otherwise.} \end{cases}$$

We call region i a *maximum (minimum)* of a signal $\mathbf{u}$ if its intensity $\bar{u}_i$ is larger (smaller) than the intensities of all its neighbors. (The term "local maximum (minimum)" would be more appropriate, but we omit the word "local" for brevity.) Region i is an *extremum* if it is either a maximum or a minimum. We let

$$\beta_i(\mathbf{u}) = \begin{cases} 1, & \text{if region } i \text{ is a maximum of } \mathbf{u}, \\ -1, & \text{if region } i \text{ is a minimum of } \mathbf{u}, \\ 0, & \text{otherwise.} \end{cases}$$

The parameter $p(\mathbf{u})$ and the four sets of parameters $n_i(\mathbf{u})$, $m_i(\mathbf{u})$, $\rho_i(\mathbf{u})$, and $\beta_i(\mathbf{u})$, for $i = 1, \ldots, p(\mathbf{u})$, are crucial to both the analysis of our 1-D algorithms and the description of their implementation. Collectively, these parameters will be referred to as *segmentation parameters* of signal $\mathbf{u}$. When it is clear from the context which signal is being described by these parameters, we will drop their arguments, and write, for example, n_i instead of $n_i(\mathbf{u})$.

The *total variation* of a signal $\mathbf{u} \in \mathbb{R}^N$ is defined by:

$$TV(\mathbf{u}) \stackrel{def}{=} \sum_{n=1}^{N-1} |u_{n+1} - u_n|,$$

and $\|\cdot\|$ stands for the usual Euclidean (ℓ^2) norm:

$$\|\mathbf{u}\|^2 \stackrel{def}{=} \sum_{n=1}^{N} u_n^2.$$

The following alternative form for $TV(\mathbf{u})$ can be obtained through a simple calculation:

$$TV(\mathbf{u}) = \sum_{i=1}^{p(\mathbf{u})} \beta_i(\mathbf{u}) \rho_i(\mathbf{u}) \bar{u}_i. \tag{2.2}$$

We are now ready to describe the 1-D SIDE which is analyzed in this paper.

2.3. A SIDE in 1-D. In [14], we provided a probabilistic interpretation for a special case of SIDEs (2.1) in 1-D. In the next section, we extend these results. The variant of 1-D SIDEs we analyze in [14] and here, is obtained by taking $F(v) = \mathrm{sgn}(v)$ (see Figure 1(b)). Specifically, we are interested in the evolution of the following system of equations:

$$\dot{u}_k(t) = \frac{1}{m_i(\mathbf{u}(t))} \left\{ \mathrm{sgn}[\bar{u}_{i+1}(t) - \bar{u}_i(t)] - \mathrm{sgn}[\bar{u}_i(t) - \bar{u}_{i-1}(t)] \right\}, \tag{2.3}$$
$$\text{for } k = n_i(\mathbf{u}(t)), n_i(\mathbf{u}(t)) + 1, \ldots, n_{i+1}(\mathbf{u}(t)) - 1,$$
$$\text{and } i = 1, \ldots, p(\mathbf{u}(t)),$$

with the initial condition:

$$\mathbf{u}(0) = \mathbf{u}^0, \tag{2.4}$$

where $\mathbf{u}^0$ is the signal to be processed. Note that when $i = 1$ and when $i = p(\mathbf{u}(t))$, Eq. (2.3) involves quantities $\bar{u}_0$ and $\bar{u}_{p(\mathbf{u}(t))+1}$ which have not been defined. We use the following conventions for these quantities:

$$\mathrm{sgn}[\bar{u}_1(t) - \bar{u}_0(t)] = \mathrm{sgn}[\bar{u}_{p+1}(t) - \bar{u}_p(t)] \stackrel{def}{=} 0. \tag{2.5}$$

Eq. (2.3) says that the intensities of samples within a region have the same dynamics, and therefore remain equal to each other. A region cannot therefore be broken into two or more regions during this evolution. The opposite, however, will happen. As soon as the intensity of some region becomes equal to that of its neighbor ($\bar{u}_j(\tau) = \bar{u}_{j+1}(\tau)$ for some j and for some time instant τ), the two become a single region–by our definition of a region. This merging of two regions into one will express itself in a change of the segmentation parameters of $\mathbf{u}(t)$ in Eq. (2.3): the number of samples in the newly formed region will be equal to the total number of samples in the two regions that are being merged; the total number of regions will be reduced by one. Borrowing our terminology from [13], we call such time instant τ when two regions get merged, a *hitting time*. Note that between two consecutive hitting times, the segmentation parameters of $\mathbf{u}(t)$ remain constant. We denote the hitting times by $t_1, \ldots, t_{p(\mathbf{u}^0)-1}$, where t_1 is the earliest hitting time and $t_{p(\mathbf{u}^0)-1}$ is the final hitting time:

$$0 < t_1 \leq t_2 \leq \ldots \leq t_{p(\mathbf{u}^0)-1}.$$

(In order to simplify notation, we also sometimes denote the final hitting time by t_f.) Note that two hitting times can be equal to each other, when more than one pair of regions are merged at the same time. Proposition 1 of the next subsection shows that the evolution (2.3,2.4) will reach a constant signal in finite time, starting from any initial condition. All samples will therefore be merged into one region, within finite time, which means that there will be exactly $p(\mathbf{u}^0) - 1$ hitting times–one fewer than the initial number of regions.

The rest of the paper considers the particular SIDE (2.3), with initial condition (2.4) and conventions (2.5). An example of its evolution, for $N = 5$, is depicted in Fig. 2.

We conclude this subsection by deriving a very useful alternative form for Eq. (2.3). If region i is a maximum of $\mathbf{u}(t)$, Eq. (2.3) and conventions (2.5) say that each of its neighbors contributes $-1/m_i$ to the rate of change of its intensity $\bar{u}_i$. Similarly, if region i is a minimum, each neighbor contributes $1/m_i$. If region i is not an extremum, then it necessarily has two neighbors, one of which contributes $1/m_i$ and the other one $-1/m_i$. In this latter case, the rate of change of $\bar{u}_i$ is zero. Combining these considerations, and using our notation from the previous subsection, we obtain an alternative form for Eq. (2.3):

$$\dot{\bar{u}}_i = -\frac{\beta_i \rho_i}{m_i}, \quad i = 1, \ldots, p, \tag{2.6}$$

where, in order to simplify notation, we did not explicitly indicate the dependence of the segmentation parameters on $\mathbf{u}(t)$.

3. Basic properties of the 1-D SIDE. In this section, we study the system (2.3), and prove a number of its properties which both allow us

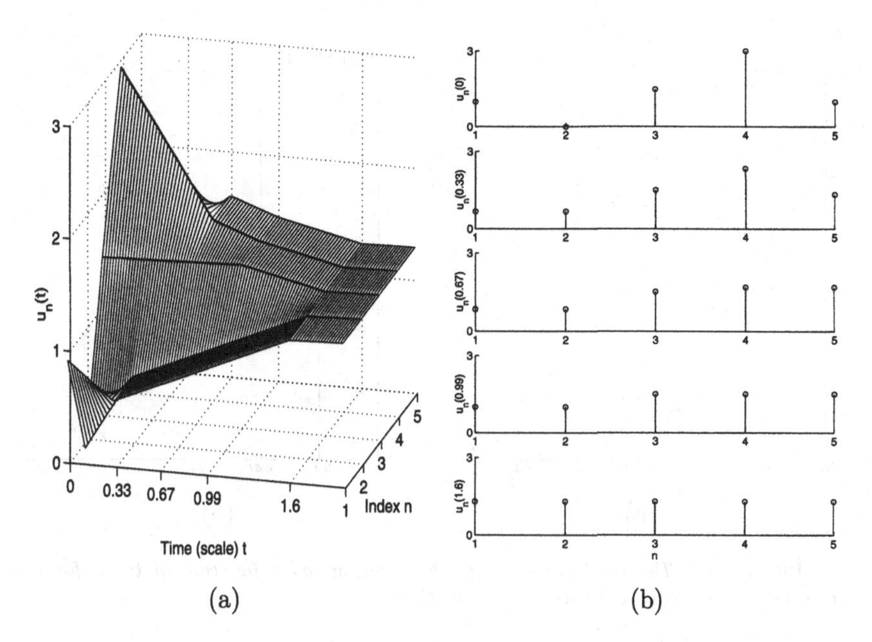

FIG. 2. *An illustration of the evolution of SIDE: (a) a mesh plot of the solution, with hitting times $t = 0.33, 0.67, 0.99, 1.6$; (b) stem plots of the initial five-point signal (top) and the solution at the four hitting times.*

to gain significant insight into its behavior, and are critical for developing optimal estimation algorithms presented in the next section. The most basic properties–illustrated in Fig. 2 and proved in [13]–assert that the system has a unique solution which is continuous, and which becomes a constant signal in finite time.

Throughout this section $\mathbf{u}(t)$ stands for the solution of (2.3,2.4), with initial condition $\mathbf{u}^0$. All the segmentation parameters encountered in this section are those of $\mathbf{u}(t)$. The final hitting time is denoted by t_f.

Proposition 1. *The solution of the SIDE (2.3,2.4) exists, is unique, is a continuous function of the time parameter t for all values of t, and is a differentiable function of t for all t except possibly the hitting times. Every pair of neighboring regions is merged in finite time. After the final hitting time t_f, the solution is a constant:*

$$u_1(t) = u_2(t) = \ldots = u_N(t) = \frac{1}{N}\sum_{n=1}^{N} u_n^0 \quad \textit{for } t \geq t_f.$$

Proof. See [13]. □

Our plan is to use the system (2.3) to solve a Gaussian estimation problem with a TV constraint, as well as a related problem originally posed in [16]. As we will see in the next section, this necessitates understanding

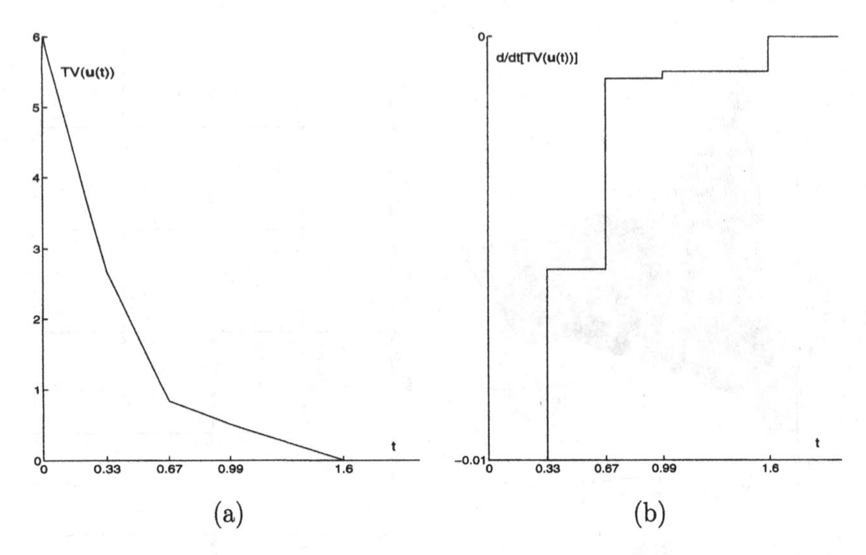

FIG. 3. *(a) The total variation of the solution, as a function of time, for the evolution of Fig. 2, and (b) its time derivative.*

the time behavior of $TV(\mathbf{u}(t))$ and -$\|\mathbf{u}(t) - \mathbf{u}^0\|^2$. Propositions 2 and 3 below show that both these quantities are Lyapunov functionals of our system (i.e., decrease as functions of time). In the same propositions, we also derive formulas for computing the time derivatives of these quantities.

Proposition 2. *For $t \in [0, t_f]$, the total variation $TV(\mathbf{u}(t))$ is a monotonically decreasing function of time, which changes continuously from $TV(\mathbf{u}(0)) = TV(\mathbf{u}^0)$ to $TV(\mathbf{u}(t_f)) = 0$ (see Fig. 3). It is a differentiable function of time except at the hitting times, and its rate of change is:*

$$\dot{TV}(\mathbf{u}(t)) = -\sum_{i=1}^{p} \frac{\beta_i^2 \rho_i^2}{m_i}. \tag{3.1}$$

Proof. Using the expression (2.2) for the TV of $\mathbf{u}(t)$, differentiating it with respect to t, and substituting (2.6) for $\dot{u}_i$ results in Eq. (3.1). This identity is valid for all $t \in [0, t_f]$ where the solution $\mathbf{u}(t)$ is differentiable and where the segmentation parameters are not changing–i.e., for all t except the hitting times. Since, by Proposition 1, the solution $\mathbf{u}(t)$ is a continuous function of t, so is $TV(\mathbf{u}(t))$. We therefore conclude that $TV(\mathbf{u}(t))$ is a monotonically decreasing continuous function of time for $t \in [0, t_f]$. Its value at t_f is zero, since, by Proposition 1, $\mathbf{u}(t_f)$ is a constant signal. □

Proposition 3. *Let $\alpha(t) = \|\mathbf{u}(t) - \mathbf{u}^0\|^2$. Then, for $t \in [0, t_f]$, $\alpha(t)$ is a monotonically increasing function of time, which changes continuously from $\alpha(0) = 0$ to $\alpha(t_f)$. It is a differentiable function of time except at the hitting times, and its rate of change is:*

$$\dot{\alpha}(t) = 2t \sum_{i=1}^{p} \frac{\beta_i^2 \rho_i^2}{m_i}. \tag{3.2}$$

Proof. The following identity is proved in [15]:

$$\sum_{k=n_i}^{n_{i+1}-1} u_k(t) = \sum_{k=n_i}^{n_{i+1}-1} u_k^0 + t(-\beta_i \rho_i). \tag{3.3}$$

We are now ready to show that Eq. (3.2) holds.

$$\begin{aligned}
\frac{d}{dt}\|\mathbf{u}(t) - \mathbf{u}^0\|^2 &= \frac{d}{dt}\sum_{n=1}^{N}(u_n(t) - u_n^0)^2 \\
&= 2\sum_{n=1}^{N}(u_n(t) - u_n^0)\dot{u}_n(t) \\
&\overset{\text{Eq. (2.3)}}{=} 2\sum_{i=1}^{p}\sum_{k=n_i}^{n_{i+1}-1}(u_k(t) - u_k^0)\left(-\frac{\beta_i \rho_i}{m_i}\right) \\
&\overset{\text{Eq. (3.3)}}{=} 2\sum_{i=1}^{p} t(-\beta_i \rho_i)\left(-\frac{\beta_i \rho_i}{m_i}\right) \\
&= 2t\sum_{i=1}^{p}\frac{\beta_i^2 \rho_i^2}{m_i}.
\end{aligned}$$

This identity is valid for all $t \in [0, t_f]$ where the solution $\mathbf{u}(t)$ is differentiable and where the segmentation parameters are not changing–i.e., for all t except the hitting times. Since, by Proposition 1, the solution $\mathbf{u}(t)$ is a continuous function of t, so is $\alpha(t)$. We therefore conclude that $\alpha(t)$ is a monotonically increasing continuous function of time for $t \in [0, t_f]$. □

Corollary 1. *Let $\alpha(t) = \|\mathbf{u}(t) - \mathbf{u}^0\|^2$, and let t_k, t_{k+1} be two consecutive hitting times. Then*

$$\alpha(t) = \alpha(t_k) + (t^2 - t_k^2)\sum_{i=1}^{p}\frac{\beta_i^2 \rho_i^2}{m_i}, \quad \textit{for any } t \in [t_k, t_{k+1}].$$

Proof. This formula is simply the result of integrating Eq. (3.2) from t_k to t, and using the fact that segmentation parameters remain constant between two hitting times. □

Proposition 4 characterizes the behavior of the functional $\|\mathbf{u}(t) - \mathbf{x}\|^2$, where $\mathbf{x}$ is an arbitrary fixed signal satisfying the constraint $TV(\mathbf{x}) \leq \nu$. This result is critical in demonstrating the optimality of our algorithms of the next section.

Proposition 4. *Let $\mathbf{u}(t)$ be the solution of (2.3,2.4), with $TV(\mathbf{u}(0)) > \nu$ and $TV(\mathbf{u}(t_\nu)) = \nu$, for some positive constants ν, t_ν. Suppose that $\mathbf{x} \in \mathbb{R}^N$*

is an arbitrary signal with $TV(\mathbf{x}) \leq \nu$. *Then, for all* $t \in [0, t_\nu]$ *except possibly the hitting times, we have:*

$$\frac{1}{2}\frac{d}{dt}\|\mathbf{u}(t) - \mathbf{x}\|^2 \leq \nu - TV(\mathbf{u}(t)),$$
$$\frac{1}{2}\frac{d}{dt}\|\mathbf{u}(t) - \mathbf{u}(t_\nu)\|^2 = \nu - TV(\mathbf{u}(t)).$$

Proof. See [15]. □

4. Optimal estimation in 1-D. In this section, we present 1-D estimation problems that can be efficiently solved using our evolution equation. We will argue in the next section that these problem formulations result not only in fast techniques for noise removal and change detection in 1-D signals, but also in 1-D prototypes of effective 2-D image processing algorithms.

4.1. ML estimation with a TV constraint. Our first example is constrained maximum likelihood (ML) estimation in additive white Gaussian noise. Specifically, suppose that the observation $\mathbf{u}^0$ is an N-dimensional vector of independent Gaussian random variables of variance σ^2, whose mean vector $\mathbf{x}$ is unknown. The only available information about $\mathbf{x}$ is that its total variation is not larger than some known threshold ν. Given the data $\mathbf{u}^0$, the objective is to produce an estimate $\hat{\mathbf{x}}$ of $\mathbf{x}$.

The ML estimate maximizes the likelihood of the observation,

$$p(\mathbf{u}^0|\mathbf{x}) = (\sqrt{2\pi}\sigma)^{-N} e^{-\frac{1}{2\sigma^2}\|\mathbf{u}^0 - \mathbf{x}\|^2}.$$

Taking into account the constraint $TV(\mathbf{x}) \leq \nu$ and simplifying the likelihood, we obtain the following optimization problem:

$$\text{Find} \quad \hat{\mathbf{x}} = \arg \min_{\mathbf{x}:TV(\mathbf{x}) \leq \nu} \|\mathbf{u}^0 - \mathbf{x}\|^2. \tag{4.1}$$

In other words, we seek the point of the constraint set $\{\mathbf{x} : TV(\mathbf{x}) \leq \nu\}$ which is the closest to the data $\mathbf{u}^0$. We now show that a fast way of solving this optimization problem is to use Eq. (2.3,2.4).

Proposition 5. *If* $TV(\mathbf{u}^0) \leq \nu$, *then the solution to Eq. (4.1) is* $\hat{\mathbf{x}} = \mathbf{u}^0$. *Otherwise, a recipe for obtaining* $\hat{\mathbf{x}}$ *is to evolve the system (2.3,2.4) forward in time until the time instant* t_ν *when the solution of the system* $\mathbf{u}(t_\nu)$ *satisfies* $TV(\mathbf{u}(t_\nu)) = \nu$. *Then* $\hat{\mathbf{x}} = \mathbf{u}(t_\nu)$. *The ML estimate is unique, and can be found in* $O(N \log N)$ *time (worst case), and with* $O(N)$ *memory requirements, where* N *is the size of the data vector.*

Discussion and a sketch of the proof. The first sentence of the proposition is trivial: if the data satisfies the constraint, then the data itself is the maximum likelihood estimate. Proposition 2 of the previous section shows that, if $\mathbf{u}(t)$ is the solution of the system (2.3,2.4), then $TV(\mathbf{u}(t))$ is a monotonically decreasing function of time, which changes continuously from $TV(\mathbf{u}^0)$ to 0 in finite time. Therefore, if $TV(\mathbf{u}^0) > \nu$, then there exists a unique time instant t_ν when the total variation of the solution is equal to ν. In the Appendix, we show that $\mathbf{u}(t_\nu)$ is indeed the ML estimate sought in Eq. (4.1), and that this estimate is unique. Finding the estimate thus amounts to solving our system of ordinary differential equations (2.3,2.4). We now sketch a fast algorithm for doing this.

Equations (2.6, 3.1) show that, between the hitting times, every intensity value $\bar{u}_i$ changes at a constant rate, and so does $TV(\mathbf{u}(t))$–as illustrated in Figs. 2 and 3. It would thus be straightforward to compute the solution once we know what the hitting times are, and which regions are merged at each hitting time.

Since a hitting time is, by definition, an instant when the intensities of two neighboring regions become equal, the hitting times are determined by the absolute values of the first differences, $v_i(t) = |\bar{u}_{i+1}(t) - \bar{u}_i(t)|$, for $i = 1, \ldots p(\mathbf{u}(t)) - 1$. Let $r_i(t) = \dot{v}_i(t)$ be the rate of change of $v_i(t)$. It follows from Eq. (2.6) that, for a fixed i, the rate $r_i(t)$ is constant between two successive hitting times:

$$v_i(t + \Delta t) = v_i(t) + \Delta t r_i(t).$$

Suppose that, after some time instant $t = \tau$, the rate $r_i(t)$ never changes: $r_i(t) = r_i(\tau)$ for $t \geq \tau$. If this were the case, we would then be able to infer from the above formula that $v_i(t)$ would become zero at the time instant $\tau + E_i(\tau)$, where $E_i(\tau)$ is defined by:

$$E_i(\tau) = \begin{cases} -v_i(\tau)/r_i(\tau), & \text{if } r_i(\tau) < 0 \\ \infty, & \text{otherwise.} \end{cases}$$

But as soon as one of the v_i's becomes equal to zero, the two corresponding regions are merged. The first hitting time is therefore $t_1 = \min_i E_i(0)$. Similarly, the second hitting time is $t_2 = t_1 + \min_i E_i(t_1)$, and, in general,

The $(k+1)$-st hitting time is $t_{k+1} = t_k + \min_i E_i(t_k)$.

To summarize, we have the following outline of the algorithm:

1. **If** $TV(\mathbf{u}(0)) \leq \nu$, **output** $\mathbf{u}(0)$ **and stop. Else, assign** $k = 0$ **and** $t_k = 0$.
2. **Find all the candidates for the next hitting time:** $t_k + E_1(t_k), \ldots, t_k + E_{p-1}(t_k)$.
3. **Store these candidates on a binary heap [8].**
4. **Find** $j = \arg\min_i [t_k + E_i(t_k)]$; **find the next hitting time** $t_{k+1} = t_k + E_j(t_k)$.
5. **If** $TV(\mathbf{u}(t_{k+1})) > \nu$, **then merge regions** j **and** $j+1$; **update** k: $k \leftarrow k+1$; **and go to Step 2.**
6. **Calculate and output** $\mathbf{u}(t_\nu)$.

The details of this algorithm are discussed in [15], where we also show that the computational complexity of the algorithm is $O(N \log N)$. Roughly speaking, the low computational cost results from Propositions 2 and 3 and Corollary 1, which provide formulas for the efficient calculation and update of the quantities of interest, as well as from the fast sorting of t_k's through the use of a binary heap [8].

4.2. The Rudin-Osher-Fatemi problem. In [16], Rudin, Osher, and Fatemi proposed to enhance images by minimizing the total variation subject to an L^2-norm constraint on the difference between the data and the estimate. In this section, we analyze the 1-D discrete[2] version of this problem:

$$\text{Find } \hat{\mathbf{x}}_{ROF} = \arg \min_{\mathbf{x}:\|\mathbf{u}^0-\mathbf{x}\|^2 \le \sigma^2} TV(\mathbf{x}), \tag{4.2}$$

where $\mathbf{u}^0$ is the signal to be processed and σ is a known parameter.

To solve this problem, we evolve our Eq. (2.3,2.4) using our algorithm of the previous subsection, but with a different stopping rule: stop at the time instant when

$$\|\mathbf{u}(t) - \mathbf{u}^0\|^2 = \sigma^2. \tag{4.3}$$

According to Proposition 3, $\|\mathbf{u}(t) - \mathbf{u}^0\|^2$ is a continuous, monotonically increasing function of time, and therefore such time instant is guaranteed to exist and be unique, as long as $0 \le \sigma^2 < \|\mathbf{u}(t_f) - \mathbf{u}^0\|^2$. Let ν be the total variation of the solution at that time instant, and denote the time instant itself by t_ν. Proposition 5 of the previous subsection says that $\mathbf{u}(t_\nu)$ is the unique solution of the following problem:

$$\min \|\mathbf{u}^0 - \mathbf{x}\|^2, \text{ subject to } TV(\mathbf{x}) \le \nu.$$

In other words, if $\mathbf{x} \neq \mathbf{u}(t_\nu)$ is any signal with $TV(\mathbf{x}) \le \nu$, then $\|\mathbf{u}^0 - \mathbf{x}\|^2 > \|\mathbf{u}^0 - \mathbf{u}(t_\nu)\|^2 = \sigma^2$. This means that if $\mathbf{x} \neq \mathbf{u}(t_\nu)$ is any signal with $\|\mathbf{u}^0 - \mathbf{x}\|^2 \le \sigma^2$, then we must have $TV(\mathbf{x}) > \nu$. Therefore, $\mathbf{u}(t_\nu)$ is the unique solution of (4.2). It is moreover shown in [15] that the new stopping rule does not change the overall computational complexity of our algorithm. We therefore have the following result.

Proposition 6. *If $\sigma^2 \ge \|\mathbf{u}(t_f) - \mathbf{u}^0\|^2$, then a solution to Eq. (4.2) which achieves zero total variation is $\hat{\mathbf{x}}_{ROF} = \mathbf{u}(t_f)$. Otherwise, the solution to Eq. (4.2) is unique, and is obtained by evolving the system (2.3,2.4) forward in time, with stopping rule (4.3). This solution can be found in $O(N \log N)$ time (worst case), and with $O(N)$ memory requirements, where N is the size of the data vector.*

[2]This means that our signals are objects in $\mathbb{R}^N$, rather than L^2.

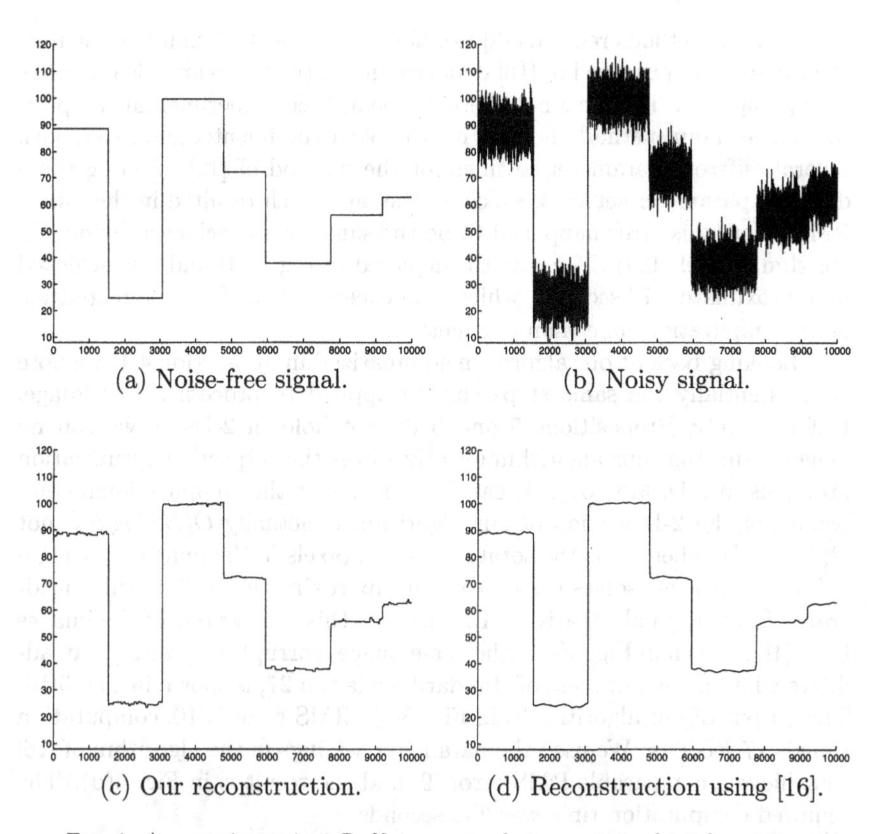

(a) Noise-free signal. (b) Noisy signal.

(c) Our reconstruction. (d) Reconstruction using [16].

FIG. 4. *An experiment in 1-D: Noise removal via constrained total variation minimization using our algorithm and algorithm in [16]. With comparable reconstruction quality and RMS errors, our algorithm takes 0.32 seconds on a SunBlade 1000, whereas the algorithm of [16], with the most favorable parameter settings, takes 72 seconds.*

5. Experimental results. In this section, we illustrate our methods with two examples. More extensive experimental results will be published in a longer paper [15].

We first apply our method of Subsection 4.2 to the Rudin-Osher-Fatemi problem in 1-D. A synthetic piecewise constant signal is shown in Fig. 4(a). The same signal, with additive zero-mean white Gaussian noise of standard deviation $\sigma = 5$, is shown in Fig. 4(b). This signal is processed using our algorithm, and the resulting output–obtained in 0.32 seconds on a SunBlade 1000–is shown in Fig. 4(c). The resulting RMS error is 0.5.

Contrasting our 1-D algorithm with others [3, 7, 9, 18], we note that, thanks to Proposition 6, our algorithm is guaranteed to achieve the solution of the Rudin-Osher-Fatemi problem *exactly* (modulo computer precision), in a finite number of steps which is asymptotically $O(N \log N)$, whereas the other methods are approximate and do not possess bounds on computational complexity. Our algorithm, moreover, relies on only one parameter,

σ; the other methods require additional parameters. For example, the iterative algorithm proposed in [16] depends on the time step and also requires a stopping rule. There cannot therefore be a direct experimental comparison between our method and that of [16]. We experimented, however, with several different parameter settings for the method of [16]. Among these different parameter settings, we chose the one which resulted in the lowest RMS error (this error happened to be the same as one achieved by our algorithm, namely 0.5). The result is depicted in Fig. 4(d) and was achieved in approximately 72 seconds, which is in excess of two orders of magnitude more computation time than our method.

Looking back at our algorithm summarized in Subsection 4.1, we note that essentially the same steps can be applied to process a 2-D image. Unfortunately, Propositions 5 and 6 do not hold in 2-D–i.e. we can no longer claim that our algorithm exactly solves the respective optimization problems in 2-D. Moreover, it can be shown that the computational complexity of the 2-D version of our algorithm is actually $O(N^2 \log N)$, not $O(N \log N)$, where N is the total number of pixels in the image. Our algorithm nevertheless solves these problems approximately and is still considerably faster than alternatives. To illustrate this, we use one of the images from [16], shown in Fig. 5(a). The same image, corrupted by zero-mean additive white Gaussian noise of standard deviation 27, is shown in Fig. 5(b). The output of our algorithm is in Fig. 5(c). RMS error is 10, computation time is 17 seconds. We used the parameter settings in the algorithm of [16] to achieve a comparable RMS error, 9, and the result is in Fig. 5(d). The required computation time was 338 seconds.

6. Conclusions. In this paper, we presented a relationship between nonlinear diffusion filtering and optimal estimation. We showed that this relationship is precise in 1-D and results in fast restoration algorithms both in 1-D and 2-D. Further results and analysis will be presented in a longer paper [15].

Acknowledgments. The author would like to thank Yan Huang for writing the code and proofreading parts of the paper.

APPENDIX

Proposition 5: Optimality proof. We need to show that

$$\mathbf{u}(t_\nu) = \arg \min_{\mathbf{x}:TV(\mathbf{x}) \le \nu} \|\mathbf{u}^0 - \mathbf{x}\|^2,$$

and that the minimum is unique. Let us denote $\phi(\mathbf{x}^1, \mathbf{x}^2) = \|\mathbf{x}^1 - \mathbf{x}^2\|^2$. To show that $\mathbf{u}(t_\nu)$ is the unique ML estimate, we therefore need to prove that

$$\phi(\mathbf{u}^0, \mathbf{u}(t_\nu)) < \phi(\mathbf{u}^0, \mathbf{x}) \text{ for any } \mathbf{x} \neq \mathbf{u}(t_\nu) \text{ such that } TV(\mathbf{x}) \le \nu.$$

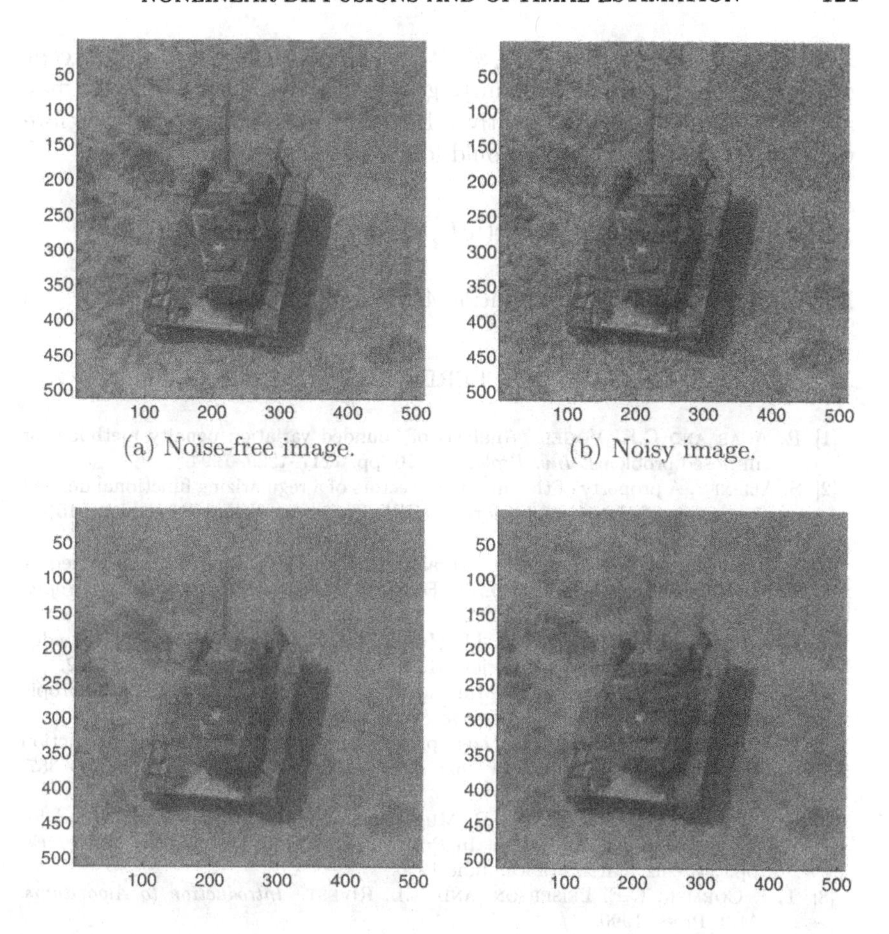

FIG. 5. *An experiment in 2-D: Noise removal via constrained total variation minimization using our algorithm and algorithm in [16]. With comparable reconstruction quality and RMS errors, our algorithm takes 17 seconds on a SunBlade 1000, whereas the algorithm of [16] takes 338 seconds.*

To compare $\phi(\mathbf{u}^0, \mathbf{u}(t_\nu))$ with $\phi(\mathbf{u}^0, \mathbf{x})$, note that

$$\phi(\mathbf{u}^0, \mathbf{u}(t_\nu)) = \phi(\mathbf{u}(t_\nu), \mathbf{u}(t_\nu)) - \int_0^{t_\nu} \frac{d}{dt} \{\phi(\mathbf{u}(t), \mathbf{u}(t_\nu))\}\, dt,$$

$$\phi(\mathbf{u}^0, \mathbf{x}) = \phi(\mathbf{u}(t_\nu), \mathbf{x}) - \int_0^{t_\nu} \frac{d}{dt} \{\phi(\mathbf{u}(t), \mathbf{x})\}\, dt.$$

Since $\phi(\mathbf{u}(t_\nu), \mathbf{u}(t_\nu)) = 0$ and $\phi(\mathbf{u}(t_\nu), \mathbf{x}) > 0$ for $\mathbf{x} \neq \mathbf{u}(t_\nu)$, our task would be accomplished if we could show that

$$\text{(A.1)} \qquad -\int_0^{t_\nu} \frac{d}{dt} \{\phi(\mathbf{u}(t), \mathbf{u}(t_\nu))\}\, dt \leq -\int_0^{t_\nu} \frac{d}{dt} \{\phi(\mathbf{u}(t), \mathbf{x})\}\, dt.$$

Note that, by Proposition 1, $\frac{d}{dt}\phi(\mathbf{u}(t), \mathbf{x})$ is well defined for all t except possibly on the finite set of the hitting times where the left derivative may not be equal to the right derivative. Both integrals in (A.1) are therefore well defined. Moreover, (A.1) would follow if we could prove that

$$\frac{d}{dt}\phi(\mathbf{u}(t), \mathbf{u}(t_\nu)) \geq \frac{d}{dt}\phi(\mathbf{u}(t), \mathbf{x}) \text{ for almost all } t \in [0, t_\nu].$$

But this is exactly what Proposition 4 of Section 2 states. □

REFERENCES

[1] R. Acar and C.R. Vogel. Analysis of bounded variation penalty methods for ill-posed problems. *Inv. Prob.*, Vol. 10, pp. 1217–1229, 1994.

[2] S. Alliney. A property of the minimum vectors of a regularizing functional defined by means of the absolute norm. *IEEE Trans. on Signal Processing*, **45**(4), April 1997.

[3] S. Alliney and S.A. Ruzinsky. An algorithm for the minimization of mixed ℓ_1 and ℓ_2 norms with applications to Bayesian estimation. *IEEE Trans. on Signal Processing*, **42**(3), March 1994.

[4] L. Alvarez, P.L. Lions, and J.-M. Morel. Image selective smoothing and edge detection by nonlinear diffusion, II. *SIAM J. Numer. Anal.*, **29**(3), 1992.

[5] M.J. Black, G. Sapiro, D.H. Marimont, and D. Heeger. Robust anisotropic diffusion. *IEEE Trans. on Image Processing*, **7**(3), 1998.

[6] C. Bouman and K. Sauer. An edge-preserving method for image reconstruction from integral projections. In *Proc. Conf. on Info. Sci. and Syst.*, pp. 383–387, Baltimore, MD, March 1991.

[7] T.F. Chan, G.H. Golub, and P. Mulet. A nonlinear primal-dual method for TV-based image restoration. In *Proc. ICAOS: Images, Wavelets, and PDEs*, pp. 241–252, Paris, France, June 1996.

[8] T.H. Cormen, C.E. Leiserson, and R.L. Rivest. *Introduction to Algorithms*, MIT Press, 1990.

[9] D. Dobson and F. Santosa. An image enhancement technique for electrical impedance tomography. *Inv. Prob.*, Vol. 10, pp. 317–334, 1994.

[10] M.G. Fleming, C. Steger, J. Zhang, J. Gao, A.B. Cognetta, I. Pollak, and C.R. Dyer. Techniques for a structural analysis of dermatoscopic imagery. *Computerized Medical Imaging and Graphics*, **22**, 1998.

[11] P. Perona and J. Malik. Scale-space and edge detection using anisotropic diffusion. *IEEE Trans. on PAMI*, **12**(7), 1990.

[12] I. Pollak. *Nonlinear Scale-Space Analysis in Image Processing.* PHD Thesis LIDS-TH-2461, Labortory for Information and Decision Systems, MIT, 1999.

[13] I. Pollak, A.S. Willsky, and H. Krim. Image segmentation and edge enhancement with stabilized inverse diffusion equations. *IEEE Trans. on Image Processing*, **9**(2), February 2000.

[14] I. Pollak, A.S. Willsky, and H. Krim. A nonlinear diffusion equation as a fast and optimal solver of edge detection problems. In *Proc. ICASSP*, Phoenix, AZ, 1999.

[15] I. Pollak, A.S. Willsky, Y. Huang, D. Mumford, and H. Krim. Nonlinear Evolution Equations as Fast and Exact Solvers of Estimation Problems. In preparation.

[16] L.I. Rudin, S. Osher, and E. Fatemi. Nonlinear total variation based noise removal algorithms. *Physica D*, 1992.

[17] P.C. Teo, G. Sapiro, and B. Wandell. Anisotropic smoothing of posterior probabilities. In *Proc. ICIP*, Santa Barbara, CA, 1997.

[18] C.R. VOGEL AND M.E. OMAN. Fast, robust total variation-based reconstruction of noisy, blurred images. *IEEE Trans. on Image Processing*, **7**(6), June 1998.

[19] S.C. ZHU AND D. MUMFORD. Prior learning and Gibbs reaction-diffusion. *IEEE Trans. on PAMI*, **19**(11), 1997.

[20] S.C. ZHU AND A. YUILLE. Region competition: unifying snakes, region growing, and Bayes/MDL for multiband image segmentation. *IEEE Trans. on PAMI*, **18**(9), 1996.

THE MUMFORD-SHAH FUNCTIONAL: FROM SEGMENTATION TO STEREO

ANTHONY YEZZI*, STEFANO SOATTO†, ANDY TSAI‡, AND ALAN WILLSKY§

Abstract. In this work, we first address the problem of simultaneous image segmentation and smoothing by approaching the Mumford-Shah paradigm from a curve evolution perspective. In particular, we let a set of deformable contours define the boundaries between regions in an image where we model the data via piecewise smooth functions and employ a gradient flow to evolve these contours.

Next, we show how a related model may be used to segment multiple images of a 3D object by evolving a surface. Projections of this surface onto each 2D image gives rise to a family of contours that define boundaries between smooth regions within each image. A key mathematical difference between this framework and the more standard Mumford-Shah framework is that the smooth function used to estimate the image data will live on the evolving interface rather than in the surrounding space.

1. Introduction. Two popular applications of partial differential equations in computer vision and image processing are found in the problems of segmentation and image smoothing. For segmentation, the technique of *snakes* or *active contours* has grown significantly since the seminal work of Kass, Witkin, and Terzopoulos [10] including the development of geometric models based on curve evolution theory [5, 6, 15, 41] and the progression from edge-based models [5, 6, 8, 10, 11, 15, 35, 36, 41] to region-based models [7, 21, 28, 42, 46, 64]. For image smoothing, the technique of anisotropic diffusion has become a widespread field of research ranging from techniques based upon the original formulation of Perona and Malik [23, 24] to curve and surface evolution methods based upon geometric scale spaces [9, 12, 13, 29] and to a number of recent techniques for color imagery and other forms of vector-valued data [33, 34, 38–40, 45].

In general, the goal of most active contour algorithms is to extract the boundaries of homogeneous regions within an image, while the goal of most anisotropic diffusion algorithms is to smooth the values of an image within homogeneous regions but not across the boundaries of such regions. We note that one of the most widely studied mathematical models in image processing and computer vision addresses both goals simultaneously, namely that of Mumford and Shah [17, 18] who presented the variational problem of minimizing a functional involving a piecewise smooth representation of an image. Their functional included a geometric term which penalized the Hausdorff measure of the set where discontinuities in the piecewise smooth estimate would be allowed. Due to the difficulties associated with

*School of Electrical and Computer Engineering, Georgia Institute of Technology, Atlanta, GA 30332.

†Department of Computer Science, UCLA, Los Angeles, CA 90095.

‡Whitaker College of Health Sciences and Technology, MIT, Cambridge, MA 02139.

§Dept. of Electrical Egineering and Computer Science, MIT, Cambridge, MA 02139.

implementing such a term in a numerical algorithm, one of the first practical numerical implementations of the Mumford-Shah model was developed by Richardson [26] and was not based upon the original functional but was based instead upon a relaxed version of the functional considered by Ambrosio and Tortorelli [3]. In this relaxed model, the exact location of boundaries between modeled homogeneous regions was "smeared" into a set with non-zero Lebesque measure, allowing the Hausdorff term to be eliminated. The recent work by Shah [31] uses the modified boundary indicator from this relaxed model as a conformal factor in a geodesic snake model, allowing the resulting algorithm to yield exact boundary locations.

In the first part of this paper, we present a curve evolution approach to minimizing the *original* Mumford-Shah functional, thereby obtaining an algorithm for simultaneous image smoothing and segmentation.[1] In contrast to anisotropic diffusion algorithms, however, the smoothing is linear, with edge preservation based upon a global segmentation as opposed to local measurements based upon the gradient. The development of this model is based upon both estimation-theoretic and geometric considerations. In particular, by viewing an active contour as the set of discontinuities considered in the original Mumford-Shah formulation, we may use the corresponding gradient flow equation to evolve the active contour. However, each gradient step involves solving an optimal estimation problem to determine piecewise smooth approximations of the image data inside and outside the active contour. We obtain these estimates by solving a linear partial differential equation (PDE) for which the solution inside the active contour is decoupled from the solution outside the active contour. This PDE, which takes the form of a Poisson equation, and the associated boundary conditions, come directly from the variational problem of minimizing the Mumford-Shah functional assuming the set of discontinuities (given by our active contour) to be fixed. The same PDE and boundary conditions can also be obtained from the theory of boundary-value stochastic processes. By taking this latter approach, we obtain an algorithm that may be regarded as a curve evolution driven by a continuum of solutions to auxiliary spatial estimation problems, connecting the theories of curve evolution and optimal estimation of stochastic processes. This development may be regarded as an extension of several recent region-based approaches to curve evolution [7, 21, 42]. In particular, it naturally generalizes the recent work of Chan and Vese in [7] who consider piecewise constant generalization of the Mumford-Shah functional within a level set framework.[2]

We note that, in general, region-based approaches enjoy a number of attractive properties including greater robustness to noise (by avoiding derivatives of the image intensity) and initial contour placement (by being

[1] A preliminary conference paper based on this work can be found in [37].

[2] The formulation in [7] can be viewed as the limiting form of our equation (1) with $\alpha = \infty$.

less local than most edge-based approaches). In contrast to most other region based techniques however (including our own previous work [42–44] and that of Chan-Vese [7] and Paragios-Deriche [21]), which assume highly constrained parametric models for pixel intensities within each region, our approach employs the statistical model directly implied by the Mumford-Shah functional. That is, the image is modeled as a random field within each region, a model that naturally accommodates variability across each region without the need to model such variability parametrically. In addition, while many region-based methods require a priori knowledge of the number of region types (such as [7] which assumes exactly two region types with two different mean intensities or [42] which requires separate sets of curves to deal with more than two region types), our Mumford-Shah based approach can automatically segment images with multiple region types (e.g. each with different mean intensities) without such a priori knowledge.

In our work, we adopt the level set techniques of Osher and Sethian [20, 30] in the implementation of our Mumford-Shah active contour model. This numerical implementation technique, in conjunction with upwind, conservative, monotone difference schemes [15, 19, 30], allows for automatic handling of cusps, corners, and topological changes as the curves evolve.

2. The Mumford-Shah formulation as a curve evolution problem. The point of reference for this paper is the Mumford-Shah functional [3]

$$E(\mathbf{f},\vec{C}) = \beta \iint_{\Omega} (\mathbf{f} - I)^2 \, dA + \alpha \iint_{\Omega \setminus \vec{C}} |\nabla \mathbf{f}|^2 \, dA + \gamma \oint_{\vec{C}} ds \tag{1}$$

in which $\vec{C}$ denotes the smooth, closed segmenting curve, I denotes the observed data, $\mathbf{f}$ denotes the piecewise smooth approximation to I with discontinuities only along $\vec{C}$, and Ω denotes the image domain [17, 18]. This energy functional is also referred to as the weak membrane by Blake and Zisserman [4]. The parameters α, β, and γ are positive real scalars which control the competition between the various terms above and determine the "scale" of the segmentation and smoothing. Of course one of these parameters can be eliminated by setting it to 1 but, for clarity of exposition, we will keep it as is. From an estimation-theoretic standpoint, the first term in $E(\mathbf{f},\vec{C})$, the data fidelity term, can be viewed as the measurement model for $\mathbf{f}$ with β inversely proportional to the variance of the observation noise process. The second term in $E(\mathbf{f},\vec{C})$, the smoothness term, can be viewed as the prior model for $\mathbf{f}$ given $\vec{C}$. The third term in $E(\mathbf{f},\vec{C})$ is a prior model

[3]The final term in the original Mumford-Shah functional consisted of a penalty on the Hausdorff measure of a more general set of discontinuities than we consider here. By restricting the discontinuity set to a smooth curve $\vec{C}$, we are able to replace this term by a simple arc length penalty.

for $\vec{C}$ which penalizes excessive arc length. With these terms, the Mumford-Shah functional elegantly captures the desired properties of segmentation and reconstruction by piecewise smooth functions. The Mumford-Shah problem is to minimize $E(\mathbf{f}, \vec{C})$ over admissible $\mathbf{f}$ and $\vec{C}$. The removal of any of the three terms in equation (1) results in trivial solutions for $\mathbf{f}$ and $\vec{C}$, yet with all three terms, it becomes a difficult problem to solve. In this paper, we constrain the set of discontinuities in the Mumford-Shah problem to correspond to evolving sets of curves, enabling us to tackle the problem via a curve-evolution-based approach.

2.1. Optimal image estimation and boundary-value stochastic processes. For any arbitrary closed curve $\vec{C}$ in the image domain, Ω is partitioned into R and R^c, corresponding to the image domain inside and outside the curve, respectively. Fixing such a curve, minimizing (1) corresponds to finding estimates $\hat{\mathbf{f}}$ and $\hat{\mathbf{g}}$ in regions R and R^c respectively, to minimize

$$
\begin{aligned}
E_{\vec{C}}(\mathbf{f}, \mathbf{g}) = \beta \iint_R (\mathbf{f} - I)^2 \, dA + \alpha \iint_R |\nabla \mathbf{f}|^2 \, dA \\
+ \beta \iint_{R^c} (\mathbf{g} - I)^2 \, dA + \alpha \iint_{R^c} |\nabla \mathbf{g}|^2 \, dA.
\end{aligned} \tag{2}
$$

The estimates $\hat{\mathbf{f}}$ and $\hat{\mathbf{g}}$ that minimize (2) satisfy (decoupled) PDE's which can be obtained using standard variational methods [17]. Alternatively, each of these estimates can be obtained from the theory of optimal estimation. This statistical interpretation is potentially of more than just intellectual interest, as it suggests lines of inquiry beyond the scope of this paper. Specifically, the estimate $\hat{\mathbf{f}}$ that minimizes (2) can be interpreted as the optimal estimate of a boundary-value stochastic process [1] $\mathbf{f}$ on the domain R whose measurement equation is

$$I = \mathbf{f} + \mathbf{v} \tag{3}$$

and whose prior probabilistic model is given by

$$\nabla \mathbf{f} = \mathbf{w} \tag{4}$$

where $\mathbf{v}$ and $\mathbf{w}$ are independent white Gaussian random fields with covariance intensities $\frac{1}{\beta}$ and $\frac{1}{\alpha}$, respectively.

One effective approach to characterizing $\mathbf{f}$ is through the use of complementary processes [1]. In particular, we seek a process $\mathbf{z}$ which *complements* the observation I in (3) in that $\mathbf{z}$ and I are uncorrelated and, together, they are informationally equivalent to $\boldsymbol{\zeta} = \{\mathbf{v}, \mathbf{w}\}$ (i.e. to all of the underlying random processes defining the estimation problem). Moreover, since the specification of the statistics of I in (3) and (4) is via a differential model and involving an internal "state" (namely $\mathbf{f}$), we seek an analogous model for $\mathbf{z}$. We refer the reader to [1] for the complete methodology for the direct construction of such complementary models, employing Green's identity and formal adjoints of differential operators. The application of this methodology to (3) and (4) yields the following model for the $\mathbf{z}$:

$$\mathbf{z} = \boldsymbol{\lambda} - \alpha \mathbf{w} \tag{5}$$

where the internal state $\boldsymbol{\lambda}$ satisfies

$$-\nabla \boldsymbol{\lambda} = \beta \mathbf{v} \tag{6}$$

with boundary condition

$$\vec{\mathcal{N}}^T \boldsymbol{\lambda} = 0 \qquad \text{on } \vec{C} \tag{7}$$

where $\mathcal{N}$ denotes the outer normal of the curve $\vec{C}$.

Eliminating $\mathbf{v}$ and $\mathbf{w}$ from (3)–(6) we can express $\mathbf{f}$ and $\boldsymbol{\lambda}$ completely in terms of I and $\mathbf{z}$. Then, since I and $\mathbf{z}$ are uncorrelated, we obtain an internal realization of the optimal estimate $\hat{\mathbf{f}}$:

$$\begin{bmatrix} \nabla & -\frac{1}{\alpha}\mathbf{I} \\ \beta \mathbf{I} & -\nabla \end{bmatrix} \begin{bmatrix} \hat{\mathbf{f}} \\ \hat{\boldsymbol{\lambda}} \end{bmatrix} = \begin{bmatrix} 0 \\ \beta I \end{bmatrix} \qquad \text{on } R$$

with the boundary condition

$$\vec{\mathcal{N}}^T \hat{\boldsymbol{\lambda}} = 0 \qquad \text{on } \vec{C}.$$

Eliminating $\hat{\boldsymbol{\lambda}}$ and noticing that the product $\vec{\mathcal{N}} \cdot \nabla \hat{\mathbf{f}}$ is the derivative of $\hat{\mathbf{f}}$ in the direction of $\vec{\mathcal{N}}$, we obtain the following damped Poisson equation with Neumann boundary condition for $\hat{\mathbf{f}}$

$$\hat{\mathbf{f}} - \frac{\alpha}{\beta} \nabla^2 \hat{\mathbf{f}} = I \qquad \text{on } R \tag{8a}$$

$$\frac{\partial \hat{\mathbf{f}}}{\partial \vec{\mathcal{N}}} = 0 \qquad \text{on } \vec{C}. \tag{8b}$$

In a similar fashion, $\hat{\mathbf{g}}$ is given as the solution to following:

$$\hat{\mathbf{g}} - \frac{\alpha}{\beta} \nabla^2 \hat{\mathbf{g}} = I \qquad \text{on } R^c \tag{9a}$$

$$\frac{\partial \hat{\mathbf{g}}}{\partial \vec{\mathcal{N}}} = 0 \qquad \text{on } \vec{C}. \tag{9b}$$

We will refer to the two sets of equations (8) and (9) as the estimation PDE's. The conjugate gradient (CG) method is employed as a fast and efficient solver for these estimation PDE's.

2.2. Gradient flows that minimize the Mumford-Shah functional. With the ability to calculate $\hat{\mathbf{f}}$ and $\hat{\mathbf{g}}$ for any given $\vec{C}$, we now wish to derive a curve evolution for $\vec{C}$ that minimizes (1). That is, as a function of $\vec{C}$, we wish to find $\vec{C}_t$ that minimizes

$$\begin{aligned} E_{\hat{\mathbf{f}},\hat{\mathbf{g}}}(\vec{C}) &= \beta \iint_R (\hat{\mathbf{f}} - I)^2 \, dA + \alpha \iint_R |\nabla \hat{\mathbf{f}}|^2 \, dA \\ &\quad + \beta \iint_{R^c} (\hat{\mathbf{g}} - I)^2 \, dA + \alpha \iint_{R^c} |\nabla \hat{\mathbf{g}}|^2 \, dA + \gamma \oint_{\vec{C}} ds. \end{aligned} \tag{10}$$

The first four terms in (10) are of the form:

$$J = \iint_D \mathcal{H} \, dA \tag{11}$$

where D denotes either the interior or the exterior of $\vec{C}$, and $\mathcal{H} : \mathbf{R}^2 \to \mathbf{R}$ is a continuous function. The gradient flow to minimize (11) is given by (see [43] for a derivation)

$$\vec{C}_t = -\mathcal{H}\vec{\mathcal{N}}. \tag{12}$$

In addition, the gradient flow that minimizes the arc length of $\vec{C}$ is given by

$$\vec{C}_t = -\mathcal{K}\vec{\mathcal{N}} \tag{13}$$

where $\mathcal{K}$ denotes the signed curvature of $\vec{C}$. Knowing gradient flows (12) and (13), the curve evolution that minimizes (10) is given by

$$\vec{C}_t = \frac{\alpha}{2}\left(|\nabla \hat{\mathbf{g}}|^2 - |\nabla \hat{\mathbf{f}}|^2\right)\vec{\mathcal{N}} + \frac{\beta}{2}\left((I - \hat{\mathbf{g}})^2 - (I - \hat{\mathbf{f}})^2\right)\vec{\mathcal{N}} - \gamma \mathcal{K}\vec{\mathcal{N}}. \tag{14}$$

For the rest of the paper, we will refer to this gradient flow as the *Mumford-Shah flow.* This flow is implemented via the level set method [20, 30] which offers a natural and numerically reliable implementation of these solutions within a context that handles topological changes in the interface without any additional effort. Furthermore this flow together with the optimal estimation PDE's makes explicit the coupling between the optimal estimates and the curve evolution.

2.3. Remarks on the Mumford-Shah active contour model. One very attractive feature associated with our Mumford-Shah active contour model (and also present in other region-based methods) is that it automatically proceeds in the correct direction without relying upon additional inflationary terms commonly employed by many active contour

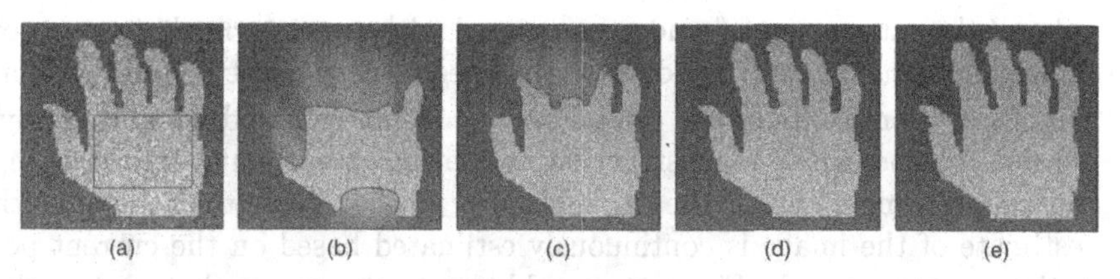

FIG. 1. *Outward flow from inside.*

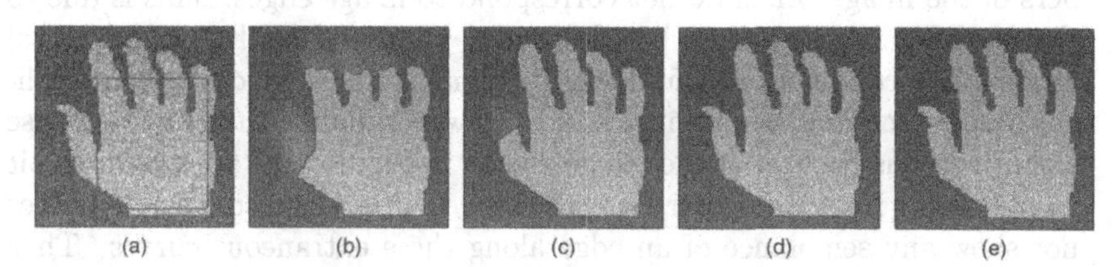

FIG. 2. *Bi-directional flow.*

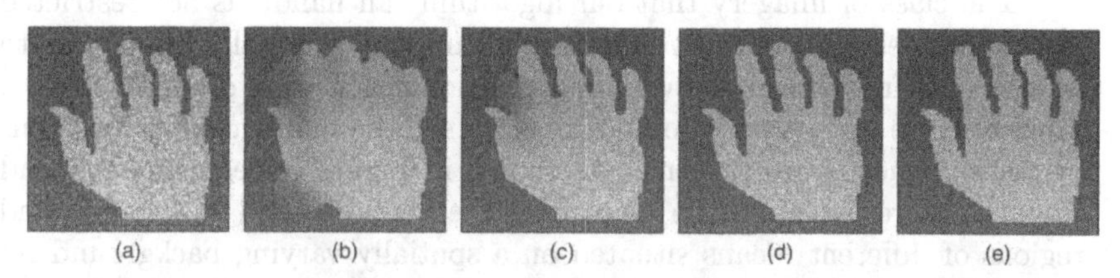

FIG. 3. *Inward flow from outside.*

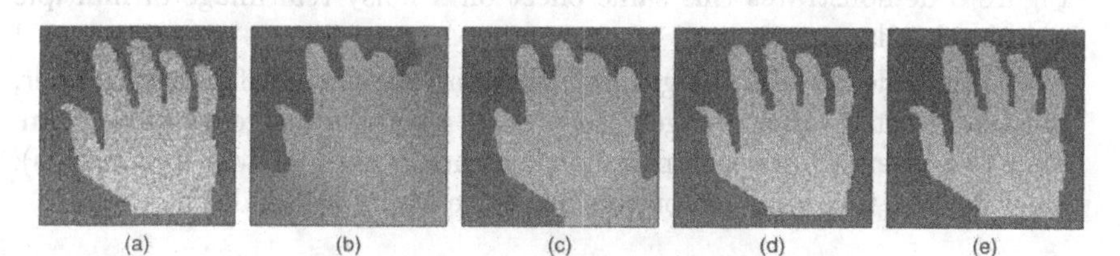

FIG. 4. *Outward flow from outside.*

algorithms. We illustrate this in Figures 1–4 with a noisy synthetic image of a hand. An initial contour completely contained within the hand will flow outward towards the boundary (Figure 1); an initial contour partially inside and partially outside the hand will flow in both directions towards the boundary (Figure 2); an initial contour encircling the hand will flow inward towards the boundary (Figure 3); and finally, an initial contour situated outside the hand will flow outward towards and wrap around the boundary (Figure 4). In these figures, Frame (a) shows the initializing contour with the original image; Frames (b) and (c) show the estimate of curve

$\vec{C}$ and the estimates of $\mathbf{f}$ and $\mathbf{g}$ associated with two intermediate steps of the algorithm; Frame (d) shows the final segmenting curve $\vec{C}$ and the final reconstruction of the image (based on the estimates $\hat{\mathbf{f}}$ and $\hat{\mathbf{g}}$); and finally, Frame (e) shows the reconstruction of the image without the overlaying curve for comparison to the original noisy image. Note that the smooth estimate of the image is continuously estimated based on the current position of the curve. In Figure 4, in addition to the curves that outline the boundary of the hand, there exist extraneous curves around the four corners of the image which do not correspond to image edges. This is due to the fact that the algorithm has descended upon and settled on to a local minimum–a common problem faced by all algorithms which rely on gradient descent methods for minimization. However, notice that the piecewise smooth reconstruction of the image shown in Figure 4(e) does not exhibit any ill effects from these extraneous curves; that is, the reconstruction does not show any semblance of an edge along these extraneous curves. Thus even if the curve is trapped at a local minimum, the reconstruction of the image is still accurate.

The class of imagery that our algorithm can handle is not restricted just to images with only two distinct means but is equally applicable to images with multiple non-overlapping regions each with different means. Moreover, we do not need to know in advance the number of such regions or distinct means are present. As shown in Figure 5, segmentation and smoothing are performed on a noisy synthetic image with four foreground regions of different means situated on a spatially varying background region. Multiple regions are captured by a single contour demonstrating the topological transitions allowed by the model's level set implementation. Figure 6 demonstrates this same effect on a noisy real image of multiple red blood cells.

Our model can also be generalized, in a very straight forward manner, to handle vector-valued images [4] (e.g. color images or images obtained from scale and orientation decompositions commonly used for texture analysis). Consider the following vector version of the Mumford-Shah functional:

$$E(\mathbf{f}_1, \mathbf{f}_2, ..., \mathbf{f}_k, \vec{C}) = \beta \iint_{\Omega} \sum_{i=1}^{k} (\mathbf{f}_i - I_i)^2 \, dA + \alpha \iint_{\Omega \setminus \vec{C}} \sum_{i=1}^{k} |\nabla \mathbf{f}_i|^2 \, dA + \gamma \oint_{\vec{C}} ds$$

where I_i and $\mathbf{f}_i$ denote the ith component of the k-dimensional vector-valued observed data and its smooth estimate, respectively. The curve evolution that minimizes this energy functional is given by

[4]Chan and Vese, who have considered the piecewise constant version of the Mumford-Shah functional [7], have also extended their framework to vector-valued data in "Active Contours without Edges for Vector-Valued Images" (see http://www.math.ucla.edu/applied/cam).

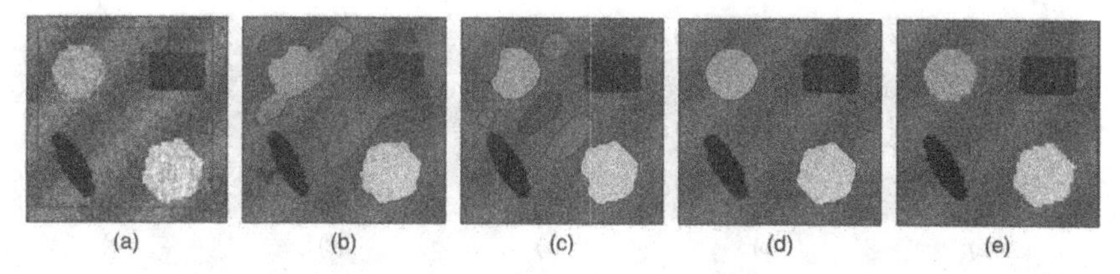

FIG. 5. *Segmentation and smoothing of an image with 4 distinct foreground regions.*

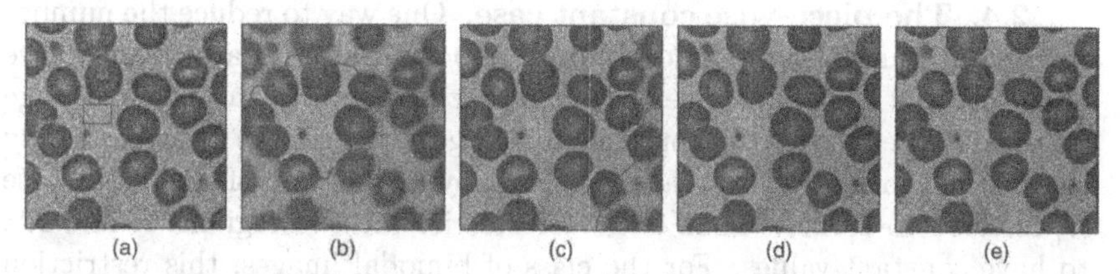

FIG. 6. *Segmentation and smoothing of an image containing multiple red blood cells.*

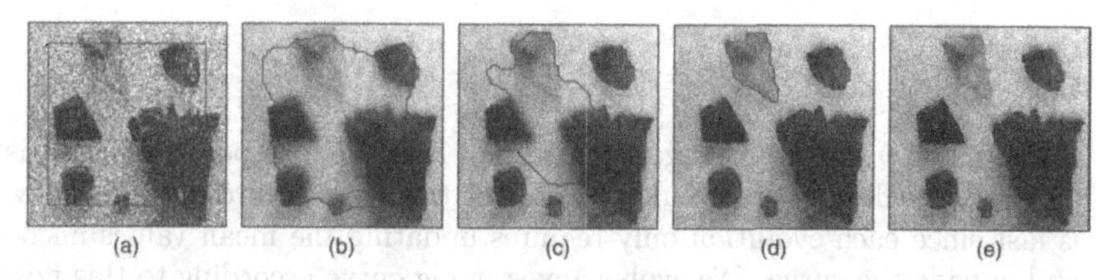

FIG. 7. *Segmentation and smoothing of a color image with 6 distinct foreground regions.*

$$(15)\quad \vec{C}_t = \frac{\alpha}{2}\sum_{i=1}^{k}\Big(|\nabla \hat{\mathbf{g}}_i|^2 - |\nabla \hat{\mathbf{f}}_i|^2\Big)\vec{\mathcal{N}} + \frac{\beta}{2}\sum_{i=1}^{k}\Big((I_i - \hat{\mathbf{g}}_i)^2 - (I_i - \hat{\mathbf{f}}_i)^2\Big)\vec{\mathcal{N}} - \gamma\mathcal{K}\vec{\mathcal{N}}.$$

The $\hat{\mathbf{f}}_i$ and $\hat{\mathbf{g}}_i$ for $i = 1, \ldots, k$ in (15) is given by the solutions to the following:

$$\hat{\mathbf{f}}_i - \frac{\alpha}{\beta}\nabla^2\hat{\mathbf{f}}_i = I_i \qquad \text{on } R$$

$$\frac{\partial \hat{\mathbf{f}}_i}{\partial \vec{\mathcal{N}}} = 0 \qquad \text{on } \vec{C}$$

and

$$\hat{\mathbf{g}}_i - \frac{\alpha}{\beta}\nabla^2\hat{\mathbf{g}}_i = I_i \qquad \text{on } R^c$$

$$\frac{\partial \hat{\mathbf{g}}_i}{\partial \vec{\mathcal{N}}} = 0 \qquad \text{on } \vec{C}.$$

For demonstration, in Figure 7, we show the segmentation and smoothing of a noisy color image of 6 different types of gemstones.

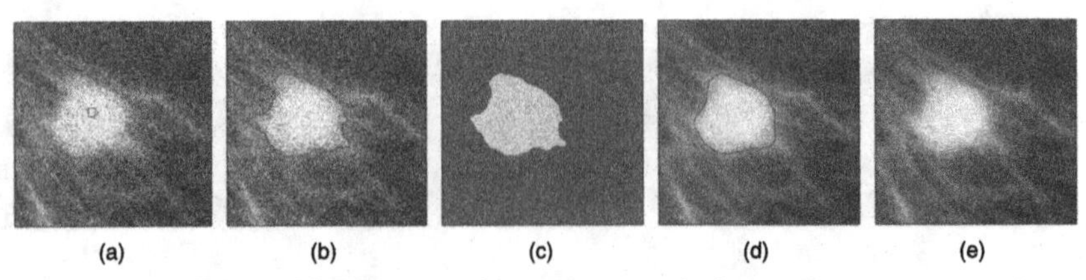

FIG. 8. *Mammogram showing cyst in the breast tissue.*

2.4. The piece-wise constant case. One way to reduce the number of curve evolution steps is to obtain a good initial estimate of the curve $\vec{C}$ so that the travel distance of the initializing curve to the correct image boundary is reduced. One approach to doing so, that works if there are only two distinct means in the image, is to employ the method of Chan and Vese [7] referred to earlier. Chan and Vese restrict the two regions, R and R^c, to have constant values. For the class of bimodal images, this restriction is equivalent to taking $\alpha = \infty$ in our Mumford-Shah active contour model. This reduces flow (14) to

$$\vec{C}_t = \frac{\beta}{2}(u - v)(I - u + I - v)\vec{\mathcal{N}} - \gamma \mathcal{K}\vec{\mathcal{N}} \tag{16}$$

where u and v are the average intensities of R and R^c, respectively. This is precisely the flow presented in [7]. Evolving the curve according to this flow is fast since each evolution only requires updating the mean values inside and outside the curve. We evolve any starting curve according to this flow in order to obtain a good initial estimate of $\vec{C}$. Once we have this estimate, we relax α to a finite value, and employ the approximate gradient descent method introduced earlier to minimize the general form of the Mumford-Shah functional. Since the initializing curve is presumably close to the correct boundary, the number of evolution steps required for convergence is greatly reduced. However, due to the use of flow (16) in calculating the initial estimate of $\vec{C}$, this 2-step implementation approach of our Mumford-Shah active contour model can only handle images containing regions with roughly two different means.

Figure 8 illustrates this 2-step implementation of our model. In Figure 8(a), a noisy mammogram showing a cyst in the breast tissue is displayed, along with the starting curve. The next frame shows the estimate of $\vec{C}$ obtained by assuming piecewise constant regions; that is, obtained by employing flow (16). This curve is superimposed on top of the original image. Figure 8(c) shows the piecewise constant approximation of the image based on this segmenting curve. In Figure 8(d) we show the results of applying the approximate gradient descent implementation of our active contour model using, as initializing curve, the one shown in Figure 8(b). Equal penalty on the arc length of the curve is use in obtaining the curves shown in Figures 8(b) and 8(d). For comparison to the original image, in Figure 8(e) we show the optimal estimate produced by our algorithm with the segmenting curve suppressed. It is clear from these results that the segmentation of the cyst has been refined and that a denoised restoration of the image is obtained.

3. Evolving 2-D curves by deforming 3-D shape. In this section we explore a modification of the paradigm of active contours, where we assume that a collection of curves in different images correspond to the occluding boundaries of an object in space. The joint evolution of the contours, therefore, corresponds to the evolution of the shape of objects in space. To begin with, we assume that a scene is composed of a number of smooth Lambertian surfaces supporting smooth Lambertian radiance functions (or dense textures with spatially homogeneous statistics). Under such assumptions, most of the significant irradiance discontinuities (or texture discontinuities) within any image of the scene correspond to occlusions between objects (or the background). These assumptions make the segmentation problem well-posed, although not general. In fact, "true" segmentation in this context corresponds directly to the shape and pose of the objects in the scene[5]. Therefore, we set up a cost functional to minimize variations within each image region, where the free parameters are not the boundaries in the image themselves, but the shape of a surface in space whose occluding contours happen to project onto such boundaries.

3.1. Notation. In what follows $\mathbf{x} = (x, y, z)$ will represent a generic point of a scene in $\Re^3$ expressed in global coordinates (based upon a fixed inertial reference frame) while $\mathbf{x}_i = (x_i, y_i, z_i)$ will represent the same point expressed in "camera coordinates" relative to an image I_i (from a sequence of images $I_1, \dots, I_n$ of the scene). To be more precise, we assume that the domain Ω_i of the image I_i belongs to a 2D plane given by $z_i = 1$ and that (x_i, y_i) constitute Cartesian coordinates within this image plane (the important point here is that the coordinate z_i corresponds to a direction which is perpendicular to the image plane). We let $\pi_i : \Re^3 \to \Omega_i; \mathbf{x} \mapsto \hat{\mathbf{x}}_i = (\hat{x}_i, \hat{y}_i)$ denote an ideal perspective projection onto this image plane, where $\hat{x}_i = x_i/z_i$ and $\hat{y}_i = y_i/z_i$. The primary objects of interest will be a regular surface S in $\Re^3$ (with area element dA) supporting a radiance function $\mathbf{f} : S \to \Re$, and a background B which we treat as a sphere of infinite radius ("blue sky") with angular coordinates $\Theta = (\theta, \phi)$ that may be related in a one-to-one manner with the coordinates $\hat{\mathbf{x}}_i$ of each image domain Ω_i through the mapping Θ_i (i.e. $\Theta = \Theta_i(\hat{\mathbf{x}}_i)$). We assume that the background supports a different radiance function $\mathbf{g} : B \to \Re$. Given the surface S, we may partion the domain Ω_i of each image I_i into a "foreground" region $R_i = \pi_i(S) \subseteq \Omega_i$, which back-projects onto the surface S, and its complement R_i^c (the "background" region), which back-projects

[5] We consider the background to be yet another object that happens to occupy the entire field of view (the "blue sky" assumption).

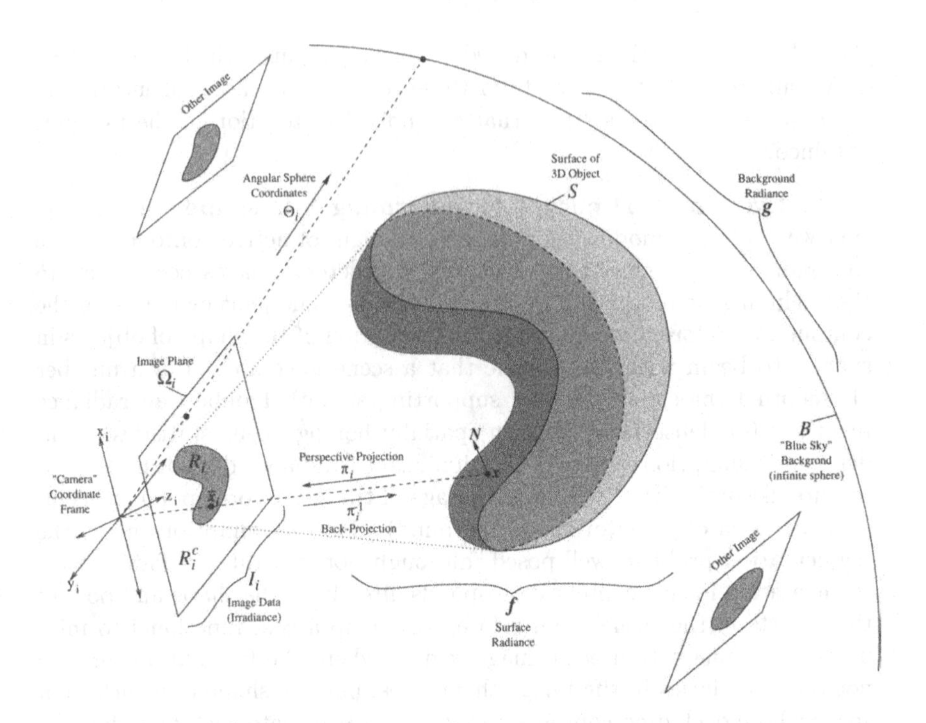

FIG. 9. *Illustration of notation used to describe the unknown surface and background (and their respective radiances), the camera images, and the relevant mappings between these objects.*

onto the background. Although the perspective projection π_i is not one-to-one (and therefore not invertible), the operation of back-projecting a point from R_i onto the surface S (by tracing back along the ray defined by $\pi_i(\text{ray}) = \hat{\mathbf{x}}_i$ until the first point on S is encountered) is indeed one-to-one with π_i as it's inverse. Therefore, we will make a slight abuse of notation and denote back-projection onto the surface S by $\pi_i^{-1} : R_i \to S$. Finally, in our computations, we will make use of the relationship between the area measure dA of the surface S and the measure $d\Omega_i = d\hat{x}_i d\hat{y}_i$ of each image domain. This arises from the form of the corresponding projection π_i and is given by $z_i^3 \, d\Omega_i = -(\mathbf{x}_i \cdot N_i) \, dA$, where N_i denotes the outward unit normal N of S expressed in the same coordinate system as $\mathbf{x}_i$. This notation is illustrated in Figure 9.

3.2. Cost functional. In order to infer the shape of a surface S, we impose a cost on the discrepancy between the projection of a model surface and the actual measurements. Such a cost, $E(\mathbf{f}, \mathbf{g}, S)$, depends upon the surface S as well as upon the radiance of the surface $\mathbf{f}$ and of the background $\mathbf{g}$. We will then adjust the model surface and radiance to match the measured images and impose a smoothness prior on radiance and a geometric prior on shape. We therefore consider the composite cost functional[6]

$$E(\mathbf{f},\mathbf{g},S) = E_{data}(\mathbf{f},\mathbf{g},S) + E_{smooth}(\mathbf{f},\mathbf{g},S) + E_{geom}(S). \tag{17}$$

We conjecture that, like in the case of the Mumford-Shah functional [17], these ingredients are sufficient to define a unique solution to the minimization problem.

In particular, the geometric and smoothness terms are given by

$$E_{geom} = \int_S dA \tag{18}$$

$$E_{smooth} = \int_S \|\nabla_S \mathbf{f}\|^2 dA + \int_B \|\nabla \mathbf{g}\|^2 d\Theta \tag{19}$$

which favor surfaces S of least surface area and radiance functions $\mathbf{f}$ and $\mathbf{g}$ of least quadratic variation. (∇_S denotes the intrinsic gradient on the surface S). Finally, the data fidelity term may be measured in the sense of $\mathcal{L}^2$ by

$$\begin{aligned} E_{data} = \sum_{i=1}^{n} \Bigg(&\int_{R_i} \big(\mathbf{f}(\pi_i^{-1}(\hat{\mathbf{x}}_i)) - I_i(\hat{\mathbf{x}}_i)\big)^2 d\Omega_i \\ &+ \int_{R_i^c} \big(\mathbf{g}(\Theta_i(\hat{\mathbf{x}}_i)) - I_i(\hat{\mathbf{x}}_i)\big)^2 d\Omega_i \Bigg). \end{aligned} \tag{20}$$

In order to facilitate the computation of the first variation with respect to S, we would rather express these integrals over the surface S as opposed to the partitions R_i and R_i^c. We start with the integrals over R_i and note that they are equivalent to

$$-\int_{\pi_i^{-1}(R_i)} \big(\mathbf{f}(\mathbf{x}) - I_i(\pi_i(\mathbf{x}))\big)^2 \frac{\mathbf{x}_i \cdot N_i}{z_i^3}\, dA = \int_{\pi_i^{-1}(R_i)} \epsilon_i^2(\mathbf{x})\sigma_i(\mathbf{x},N)\, dA$$

where $\epsilon_i(\mathbf{x}) = \mathbf{f}(\mathbf{x}) - I_i(\pi_i(\mathbf{x}))$ and $\sigma_i(\mathbf{x},N) = -(\mathbf{x}_i \cdot N_i)/z_i^3$. Now we move to the integrals over R_i^c and note that they are equivalent to

$$\int_{\Omega_i} \varepsilon_i^2(\hat{\mathbf{x}}_i)\, d\Omega_i - \int_{\pi_i^{-1}(R_i)} \varepsilon_i^2(\pi_i(\mathbf{x}))\sigma_i(\mathbf{x},N)\, dA$$

where $\varepsilon_i(\hat{\mathbf{x}}_i) = \mathbf{g}\big(\Theta_i(\hat{\mathbf{x}}_i) - I_i(\hat{\mathbf{x}}_i)\big)$. Combining these "restructured" integrals yields:

$$\int_{\Omega_i} \varepsilon_i^2(\hat{\mathbf{x}}_i)\, d\Omega_i + \int_{\pi_i^{-1}(R_i)} \big(\epsilon_i^2(\mathbf{x}) - \varepsilon_i^2(\pi_i(\mathbf{x}))\big)\sigma_i(\mathbf{x},N)\, dA$$

[6] Again we could consider each term weighted by a coefficient; however, for simplicity of notation, we let such weights to be equal to one.

Note that the first integral in the above expression is independent of the surface S (and its radiance function $\mathbf{f}$) and that the second integral is taken over only a subset of S given by $\pi_i^{-1}(R_i)$. We may express this as an integral over all of S (and thereby avoid the use of π_i^{-1} in our expression) by introducing a characteristic function $\chi_i(\mathbf{x}) \in \{0,1\}$ into the integrand where $\chi_i(\mathbf{x}) = 1$ for $\mathbf{x} \in \pi_i^{-1}(R_i)$ (represented by the shaded part of the surface S in Figure 9) and $\chi_i(\mathbf{x}) = 0$ for $\mathbf{x} \notin \pi_i^{-1}(R_i)$ (i.e. for points that are occluded by other points on S). We therefore obtain the following equivalent expression for E_{data} given in (20):

$$(21)\quad E_{data} = \sum_{i=1}^{n}\left(\int_{\Omega_i}\varepsilon_i^2(\hat{\mathbf{x}}_i)d\Omega_i + \int_S \chi_i(\mathbf{x})\big(\epsilon_i^2(\mathbf{x}) - \varepsilon_i^2(\pi_i(\mathbf{x}))\big)\sigma_i(\mathbf{x},N)dA\right).$$

Finally, we express the entire cost functional by adding the geometric and smoothness priors:

$$E(\mathbf{f},\mathbf{g},S) = \sum_{i=1}^{n}\left(\int_{\Omega_i}\varepsilon_i^2(\hat{\mathbf{x}}_i)\,d\Omega_i + \int_S \chi_i(\mathbf{x})\big(\epsilon_i^2(\mathbf{x}) - \varepsilon_i^2(\pi_i(\mathbf{x}))\big)\sigma_i(\mathbf{x},N)\,dA\right) + \int_S \|\nabla_S \mathbf{f}\|^2 dA + \int_B \|\nabla \mathbf{g}\|^2 d\Theta + \int_S dA.$$

3.3. Evolution equation. The variation of E_{geom}, which is just the surface area of S, is given by

$$-\frac{d}{dS}E_{geom} = -HN,$$

where H denotes mean curvature and N the outward unit normal. The variation of E_{smooth} is given by

$$-\frac{d}{dS}E_{smooth} = \left(K\left\langle \nabla_S\mathbf{f}, A^{-1}\nabla_S\mathbf{f}\right\rangle - \|\nabla_S\mathbf{f}\|H\right)N,$$

where K denotes the Gaussian curvature of S, where ∇_S denotes the gradient of $\mathbf{f}$ taken with respect to isothermal coordinates (the "intrinsic gradient" on S), and where A denotes the second fundamental form of S with respect to these coordinates.

The variation of E_{data} requires some attention. In fact, the data fidelity term in (21) involves an explicit model of occlusions[7] via a characteristic function. Discontinuities in the kernel cause major problems, for they can result in variations that are zero almost everywhere (e.g. for the case of constant radiance). One easy solution is to mollify the corresponding gradient flow. This can be done in a mathematically sound way by interpolating a smooth force field on the surface in space. Alternatively,

[7] The geometric term and the smoothness term are independent of occlusions.

the characteristic functions χ_i in the data fidelity term can be mollified, thereby making the integrands differentiable everywhere.

In order to arrive at an evolution equation, we note that the components of the data fidelity term, as expressed in equation (21), that depend upon S have the following form

$$E_i(S) = \int_S G_i(\mathbf{x}) \cdot N_i \, dA. \tag{22}$$

The gradient flows corresponding to such energy terms have the form

$$-\frac{d}{dS} E_i = -(\nabla_i \cdot G_i) N, \tag{23}$$

where ∇_i denotes the gradient with respect to $\mathbf{x}_i$ (recall that $\mathbf{x}_i$ is the representation of a point using the camera coordinates associated with image I_i as described in Section 3.1). In particular,

$$G_i(\mathbf{x}) = -\chi_i(\mathbf{x}) \left(\epsilon_i^2(\mathbf{x}) - \varepsilon_i^2(\pi_i(\mathbf{x}))\right) \frac{\mathbf{x}_i}{z_i^3} \tag{24}$$

and the divergence of G_i, after simplification, is given by

$$\begin{aligned} -\nabla_i \cdot G_i = \frac{1}{z_i^3} \Big((\mathbf{f} - \mathbf{g}) \Big[(I_i - \mathbf{f}) + (I_i - \mathbf{g}) \Big] (\nabla_i \chi_i \cdot \mathbf{x}_i) \\ + 2\chi_i (I_i - \mathbf{f})(\nabla_i \mathbf{f} \cdot \mathbf{x}_i) \Big), \end{aligned} \tag{25}$$

where we have omitted the arguments of $\mathbf{f}$, $\mathbf{g}$, and I_i for the sake of simplicity. A particularly nice feature of this final expression (which is shared by the standard Mumford-Shah formulation for direct image segmentation) is that it depends only upon the image values, *not upon the image gradient*, which makes it less sensitive to image noise when compared to other variational approaches to stereo (and therefore less likely to cause the resulting flow to become "trapped" in local minima). Notice that the first term in this flow involves the gradient of the characteristic function χ_i and is therefore non-zero only on the portions of S which project (π_i) onto the *boundary* of the region R_i (illustrated by the dashed curve on the surface S in Figure 9). As such, this term may be directly associated with a curve evolution equation for the boundary of the region R_i within the domain Ω_i of the image I_i. The second term, on the other-hand, may be non-zero over the entire patch $\pi_i^{-1}(R_i)$ of S (illustrated by entire the shaded part of the surface S in Figure 9).

We may now write down the complete gradient flow for $E = E_{data} + E_{smooth} + E_{geom}$ as

$$\begin{aligned} \frac{dS}{dt} = \sum_{i=1}^{n} \frac{1}{z_i^3} \Big((\mathbf{f} - \mathbf{g}) \Big[(I_i - \mathbf{f}) + (I_i - \mathbf{g}) \Big] (\nabla_i \chi_i \cdot \mathbf{x}_i) \\ + 2\chi_i (I_i - \mathbf{f})(\nabla_i \mathbf{f} \cdot \mathbf{x}_i) \Big) N + \left(K \left\langle \nabla_S \mathbf{f}, A^{-1} \nabla_S \mathbf{f} \right\rangle - \|\nabla_S \mathbf{f}\| H \right) N - HN. \end{aligned} \tag{26}$$

3.4. Estimating scene radiance. Once an estimate of the surface S is available, the radiance estimates $\mathbf{f}$ and $\mathbf{g}$ must be updated. For a given surface S we may regard our energy functional $E(S, \mathbf{f}, \mathbf{g})$ as a function only of $\mathbf{f}$ and $\mathbf{g}$ and minimize it accordingly. A necessary condition is that $\mathbf{f}$ and $\mathbf{g}$ satisfy the Euler-Lagrange equations for E based upon the current surface S. These optimal estimate equations are given by the following elliptic PDE's on the surface S and the background B,

$$\Delta_S \mathbf{f} = \sum_{i=1}^{n} \chi_i(\mathbf{f} - I_i)\sigma_i \quad \text{and} \quad \Delta_\Theta \mathbf{g} = \sum_{i=1}^{n} \hat{\chi}_i(\mathbf{g} - I_i) \tag{27}$$

where Δ_S denotes the Laplace-Beltrami operator on the surface S, where Δ_Θ denotes the Laplacian on the background B with respect to its spherical coordinates Θ, and where $\hat{\chi}_i(\Theta)$ denotes a characteristic function for the background B where $\hat{\chi}_i(\Theta)$=1 if $\Theta_i^{-1}(\Theta) \in R_i^c$ and $\hat{\chi}_i(\Theta)$=0 otherwise.

3.5. The piecewise constant case. We obtain a special piecewise constant[8] energy functional as a limiting case of the more general energy functional (17) by giving the smoothness term E_{smooth} infinite weight. In this case, the only critical points are constant radiance functions. We may obtain an equivalent formulation, by dropping the E_{smooth} term

$$\begin{aligned} E_{constant} &= E_{data} + E_{smooth} \\ &= \sum_{i=1}^{n} \left(\int_{R_i} \left(\mathbf{f}(\pi_i^{-1}(\hat{\mathbf{x}}_i)) - I_i(\hat{\mathbf{x}}_i)\right)^2 d\Omega_i \right. \\ &\quad \left. + \int_{R_i^c} \left(\mathbf{g}(\Theta_i(\hat{\mathbf{x}}_i)) - I_i(\hat{\mathbf{x}}_i)\right)^2 d\Omega_i \right) + \int_S dA, \end{aligned} \tag{28}$$

and by restricting our class of admissible radiance functions $\mathbf{f}$ and $\mathbf{g}$ to be only constants. In this simpler formulation, one no longer needs to solve a PDE on the surface S nor on the background B, to obtain optimal estimates for $\mathbf{f}$ and $\mathbf{g}$ (given the current location of the surface S). In this case, $E_{constant}$ is minimized by setting the constants $\mathbf{f}$ and $\mathbf{g}$ to be the overall sample mean of I_i over the regions R_i (for each $1 \leq i \leq n$) and the overall sample mean of I_i over the complementary regions R_i^c respectively. The gradient flow associated with the E_{data} term simplifies. Recall that, in the general case, the E_{data} gradient flow depends upon two terms (given by (25)), one of which only acts upon the points of S which project to the boundaries of the regions R_i, giving rise to curve evolutions for these segmentation boundaries, while the second term acts upon each entire patch

[8]We say "piecewise constant" even though each radiance function is treated as a single constant since the segmentations obtained by projecting these objects with constant radiances onto the camera images yield *piecewise constant* approximations to the image data.

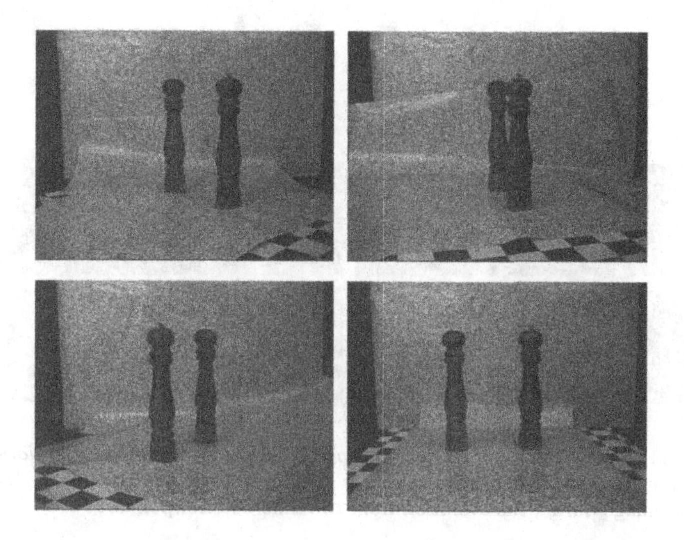

FIG. 10. *Original "salt and pepper" sequence (4 of 22 views).*

of S associated with each region R_i. In the piecewise constant case, this second term drops out (since it depends upon the gradient of $\mathbf{f}$), and therefore only the boundary evolution term remains. As such, the gradient flow for $E_{constant}$ is given by

$$\frac{dS}{dt} = \Big(\sum_{i=1}^{n} \frac{1}{z_i^3}(\mathbf{f}-\mathbf{g})\Big[(I_i-\mathbf{f})+(I_i-\mathbf{g})\Big](\nabla_i\chi_i\cdot\mathbf{x}_i)N\Big) - HN. \tag{29}$$

A numerical implementation of the evolution equation above has been carried out within the level set framework of Osher and Sethian [20]. A number of sequences has been captured and the relative position and orientation of each camera has been computed using standard camera calibration methods. Here we show the results on two representative experiments. Although the equation above assumes that the scene is populated by objects with constant albedo, the reader will recognize that the scenes we have tested our algorithms on represent significant departures from such assumptions. They include fine textures, specular highlights and even substantial calibration errors.

In Figure 10 we show 4 of 22 calibrated views of a scene that contains three objects: two shakers and the background. This scene would represent an insurmountable challenge to traditional correspondence-based stereo algorithms: the shakers exhibit very little texture (making local correspondence ill-posed), while the background exhibits very dense texture (making local correspondence prone to local minima). In addition, the shakers have a dark but shiny surface, that reflects highlights that move

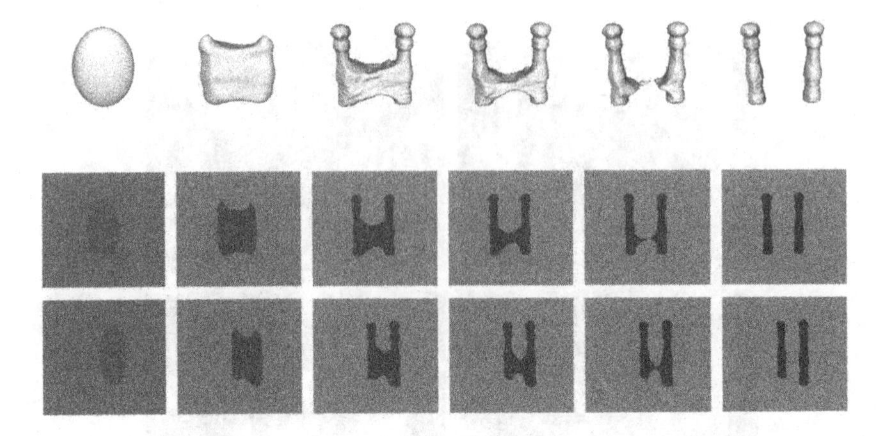

FIG. 11. *(Top) Rendered surface during evolution (6 of 800 steps). Notice that the initial surface is neither contains nor is contained by the final surface. (Bottom) segmented image during the evolution from two different viewpoints.*

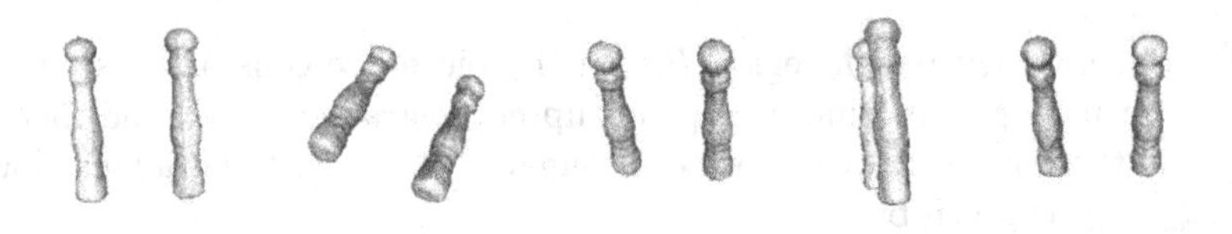

FIG. 12. *Final estimated surface shown from several viewpoints. Notice that the bottoms of the salt and pepper shakers are flat, even though no data was available. This is due to the geometric prior, which in the absence of data results in a minimal surface being computed.*

relative to the camera since the scene is rotated while the light is kept stationary. In Figure 11 we show the surface evolving from a large ellipse that neither contains nor is contained in the shape of the scene, to a final solid model. Notice that some parts of the initial surface evolve outward, while other parts evolve inward in order to converge to the final shape. This bi-directionality is a feature of our algorithm, which is not shared - for instance - by shape carving methodologies. There, once a pixel has been deleted, it cannot be retrieved.

In Figure 12 we show the final result from various vantage points. In Figure 13 we show the final segmentation in some of the original views (top). We also show the segmented foreground superimposed to the original images. Two of the 22 views were poorly calibrated, as it can be seen from the large reprojection error. However, this does not significantly impact the final reconstruction, for there is an averaging effect by integrating data from all views.

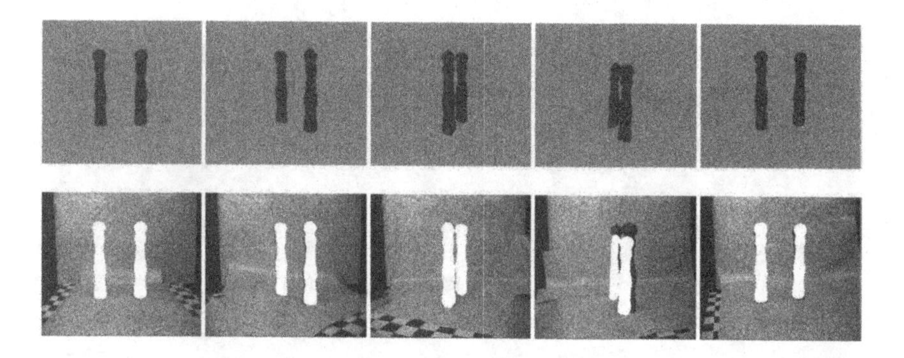

FIG. 13. *(Top) Image segmentation for the salt and pepper sequence. (Bottom) Segmented foreground superimposed to the original sequence. The calibration in two of the 22 images was inaccurate (one of which is shown above; the fourth image from the left). However, the effect is mitigated by the global integration, and the overall shape is only marginally affected by the calibration errors.*

FIG. 14. *The "watering can" sequence and the initial surface. Notice that the initial surface is not simply connected and neither contains nor is contained by the final surface. In order to capture a hole it is necessary that it is intersected by the initial surface. One way to guarantee this is to start with a number of small surfaces.*

In Figure 14 we show an image from a sequence of views of a watering can, together with the initial surface. The estimated shape is shown in Figure 15.

Acknowledgements. This research is supported in part by NSF grant IIS-9876145, ARO grant DAAD19-99-1-0139 and Intel grant 8029. We thank Wilfrid Gangbo and Luigi Ambrosio for very helpful discussions and insights, Allen Tannenbaum and Guillermo Sapiro for comments and suggestions. We also wish to thank Luminita Vese, Tony Chan and Stan Osher for comments.

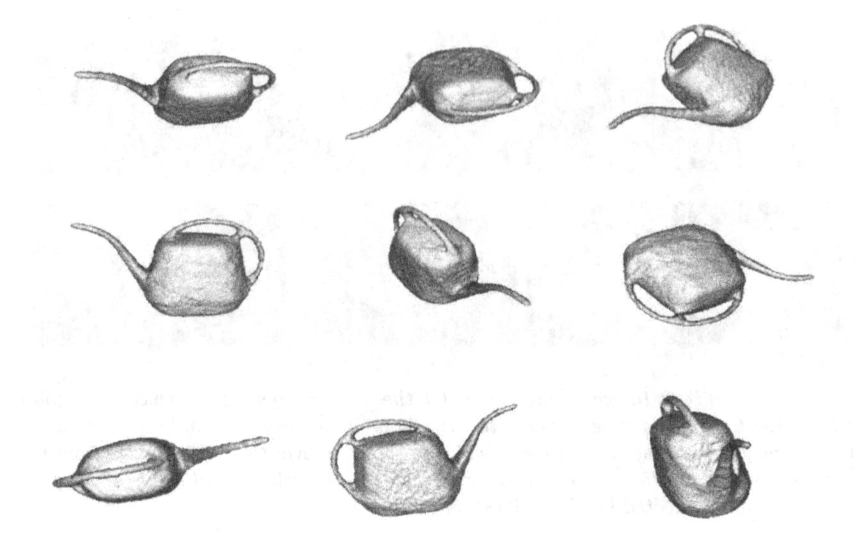

FIG. 15. *Final estimated shape for the watering can. The two initial surfaces, as seen in Figure 14, have merged. Although no ground truth is available for these sequences, it is evident that the topology and geometry of the watering can has been correctly captured.*

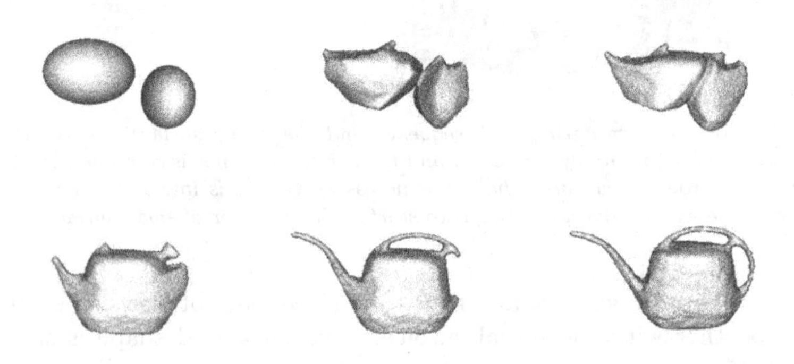

FIG. 16. *(top) Rendered surface during evolution for the watering can.*

REFERENCES

[1] M. Adams, A. Willsky, and B. Levy, "Linear estimation of boundary value stochastic processes–part 1: The role and construction of complementary models," *IEEE Trans. Automatic Control*, **29**: 803–811, 1984.

[2] J. Allebach and W. Wong, "Edge directed interpolation," *IEEE International Conference on Image Processing*, **3**: 707–711, 1996.

[3] L. Ambrosio and V.M. Tortorelli, "Approximation of functionals depending on jumps by elliptic functionals via Γ-convergence," *Communications in Pure and Applied Math*, **43**(8), 1990.

[4] A. Blake and A. Zisserman, *Visual Reconstruction*, MIT Press, 1987.

[5] V. Caselles, F. Catte, T. Coll, and F. Dibos, "A geometric model for active contours in image processing," *Numerische Mathematik*, **66**: 1–31, 1993.

[6] V. Caselles, R. Kimmel, and G. Sapiro, "Geodesic snakes," *Int. J. Computer Vision*, 1998.

[7] T. Chan and L. Vese, "Active contours without edges," UCLA Technical Report, submitted to *IEEE Trans. Image Processing*.

[8] L. Cohen, "On active contour models and balloons," *CVGIP: Image Understanding*, **53**: 211–218, 1991.

[9] A.I. El-Fallah and G.E. Ford, "Mean curvature evolution and surface area scaling in image filtering," *IEEE Trans. on Image Processing*, **6**: 750–753, 1997.

[10] M. Kass, A. Witkin, and D. Terzopoulos, "Snakes: active contour models," *Int. Journal of Computer Vision*, **1**: 321–331, 1987.

[11] S. Kichenassamy, A. Kumar, P. Olver, A. Tannenbaum, and A. Yezzi, "Conformal curvature flows: from phase transitions to active vision," *Arch. Rational Mech. Anal.*, **134**: 275–301, 1996.

[12] B.B. Kimia and K. Siddiqi, "Geometric heat equation and nonlinear diffusion of shapes and images," *Proc. IEEE Conf. Computer Vision and Pattern Recognition*, pp. 113–120, 1994.

[13] R. Kimmel, "Affine differential signatures for gray level images of planar shapes," *Int. Conf. on Computer Vision*, 1996.

[14] J. Lim, *Two-dimensional Signal and Image Processing*, Prentice Hall, 1992.

[15] R. Malladi, J. Sethian, and B. Vemuri, "Shape modeling with front propagation: a level set approach," *IEEE Trans. Pattern Anal. Machine Intell.*, **17**: 158–175, 1995.

[16] B. Merriman, J. Bence, and S. Osher, "Motion of multiple junctions: A level set approach," *Journal of Computational Physics*, **112**(2): 334–363, 1994.

[17] D. Mumford and J. Shah, "Optimal approximations by piecewise smooth functions and associated variational problems," *Communications in Pure and Applied Mathematics*, **42**(4), 1989.

[18] D. Mumford and J. Shah, "Boundary detection by minimizing functionals," *Proc. IEEE Conf. Computer Vision and Pattern Recognition*, 1985.

[19] S. Osher, "Riemann solvers, the entropy condition, and difference approximations," *SIAM J. Numer. Anal.*, **21**: 217–235, 1984.

[20] S. Osher and J. Sethian, "Fronts propagation with curvature dependent speed: Algorithms based on Hamilton-Jacobi formulations," *Journal of Computational Physics*, **79**: 12–49, 1988.

[21] N. Paragios and R. Deriche, "Geodesic active regions for texture segmentation," Research Report 3440, INRIA, France, 1998.

[22] P. Perona, "Orientation diffusions," *IEEE Trans on Image Processing*, **7**: 457–467, 1998.

[23] P. Perona and J. Malik, "Scale-space and edge detection using anisotropic diffusion," *IEEE Trans. on Pattern Anal. Machine Intell.*, **12**: 629–639, 1990.

[24] P. Perona and T. Shiota and J. Malik, "Anisotropic diffusion." In [27].

[25] K. Ratakonda and N. Ahuja, "POC based adaptive image magnification," *IEEE International Conference on Image Processing.*, **3**: 203–207, 1998.

[26] T.J. Richardson, *Scale independent piecewise smooth segmentation of images via variational methods.* Ph.D. Thesis, Dept. of E.E.C.S., M.I.T., 1989.

[27] B.M. ter Harr Romeny, editor. *Geometry-Driven Diffusion in Computer Vision.* Kluwer Academic Publishers, 1994.

[28] R. Ronfard, "Region-based strategies for active contour models," *Int. J. Computer Vision*, **13**: 229–251, 1994.

[29] G. Sapiro and A. Tannenbaum, "Area and length preserving geometric invariant scale-space," *IEEE Trans. on Pattern Anal. Machine Intell.*, **17**: 67–72, 1995.

[30] J. Sethian, *Level Set Methods: Evolving Interfaces in Geometry, Fluid Mechanics, Computer Vision, and Material Science*, Cambridge Univ. Press, 1996.

[31] J. Shah, "A common framework for curve evolution, segmentation and anisotropic diffusion," *Proc. IEEE Conf. Computer Vision and Pattern Recognition*, 1996.

[32] K. Siddiqi, Y. Lauziere, A. Tannenbaum, and S. Zucker, "Area and length minimizing flows for segmentation," *IEEE Trans. Image Processing*, **7**: 433–444, 1998.

[33] G. Sapiro and D.L. Ringarch, "Anisotropic diffusion of multivalued images with applications to color filtering," *IEEE Trans. on Image Processing*, **5**: 1582–1585, 1996.

[34] N. Sochen and R. Kimmel and R. Malladi, "A general framework for low level vision," *IEEE Trans. on Image Processing*, **7**: 310–318, 1998.

[35] H. Tek and B. Kimia, "Image segmentation by reaction diffusion bubbles," *Proc. Int. Conf. Computer Vision*, pp. 156–162, 1995.

[36] D. Terzopoulos and A. Witkin, "Constraints on deformable models: recovering shape and non-rigid motion," *Artificial Intelligence*, **36**: 91–123, 1988.

[37] A. Tsai, A. Yezzi, and A. Willsky, "A curve evolution approach to smoothing and segmentation using the Mumford-Shah functional," *Proc. IEEE Conf. Computer Vision and Pattern Recognition*, 2000, to appear.

[38] J. Weickert, "Scale-space properties of nonlinear diffusion filtering with a diffusion tensor," Tech. Rep.110, Lab. Technomath., Univ. Kaiserslautern, Germany, 1994.

[39] R. Whitaker and G. Gerig, "Vector-valued diffusion," *Geometry Driven Diffusion in Computer Vision*, B. ter Haar Romeny, Ed. Boston, MA: Kluwer, 1994, pp. 93–133.

[40] A. Yezzi, "Modified curvature motion for image smoothing and enhancement," *IEEE Trans. on Image Processing*, **7**: 345–352, 1998.

[41] A. Yezzi, S. Kichenassamy, A. Kumar, P. Olver, and A. Tannenbaum, "A geometric snake model for segmentation of medical imagery," *IEEE Trans. Medical Imaging*, **16**: 199–209, 1997.

[42] A. Yezzi, A. Tsai, and A. Willsky, "A statistical approach to snakes for bimodal and trimodal imagery" *Int. Conf. on Computer Vision*, 1999.

[43] A. Yezzi, A. Tsai, and A. Willsky, "Fully global, coupled curve evolution equations for image segmentation" Submitted to *IEEE Trans. on Pattern Anal. Machine Intell.*

[44] A. Yezzi, A. Tsai, and A. Willsky, "A fully global approach to image segmentation via coupled curve evolution equations," Submitted to *Journal of Visual Communication and Image Representation*

[45] Y.L. You and W. Zhu and A. Tannenbaum and M. Kaveh, "Behavioral analysis of anisotropic diffusion," *IEEE Trans. Image Processing*, **5**: 1539–1553

[46] S. C. Zhu, T. S. Lee, and A. L. Yuille, "Region Competition: Unifying snakes, region growing, and Bayes/MDL for multiband image segmentation," *Proc. Int. Conf. Computer Vision*, pp. 416–423, 1995.

[47] M. Bertalmio, L. Cheng, S. Osher, and G. Sapiro, "Variational Problems and Partial Differential Equations on Implicit Surfaces: The Framework and Examples in Image Processing and Pattern Formation". *CAM Technical Report 00-23*, UCLA, June 2000.

[48] A. Chakraborty and J. Duncan, "Game-Theoretic Integration for Image Segmentation," *IEEE Trans. Pattern Anal. Machine Intell.*, **21**(1): 12–30, Jan. 1999.

[49] A. Chakraborty, L. Staib, and J. Duncan, "Deformable Boundary Finding in Medical Images by Integrating Gradient and Region Information," *IEEE Trans. Medical Imaging*, **15**(6): 859–870, Dec. 1996.

[50] R. Cipolla and A. Blake. Surface shape from the deformation of apparent contours. *Int. J. of Computer Vision, 9 (2)*, 1992.

[51] M. Crandall, H. Ishii, and P. Lions, "Users guide to viscosity solutions of second order partial differential equations," *Bulletin of Amer. Math. Soc.*, **27**: 1–67, 1992.

[52] O. Faugeras. *Three dimensional vision, a geometric viewpoint.* MIT Press, 1993.

[53] O. Faugeras and R. Keriven. Variational principles, surface evolution pdes, level set methods and the stereo problem. *INRIA Technical report*, 3021:1–37, 1996.

[54] W. Fleming and H. Soner, *Controlled Markov processes and viscosity solutions.* Springer-Verlag, New York, 1993.

[55] G. Gimelfarb and R. Haralick, "Terrain reconstruction from multiple views," *in PRoc. 7th Intl. Conf. on Computer Analysis of Images and Patterns,* pp. 695–701, 1997.

[56] B. Horn and M. Brooks (eds.). *Shape from Shading.* MIT Press, 1989.

[57] K. Kutulakos and S. Seitz. A theory of shape by space carving. In *Proc. of the Intl. Conf. on Comp. Vision*, 1998.

[58] Y. Leclerc, "Constructing stable descriptions for image partitioning," *Int. J. Computer Vision*, **3**: 73–102, 1989.

[59] R. J. LeVeque, *Numerical Methods for Conservation Laws*, Birkhäuser, Boston, 1992.

[60] P. L. Lions, *Generalized Solutions of Hamilton-Jacobi Equations*, Pitman Publishing, Boston, 1982.

[61] N. Paragios and R. Deriche, "Coupled Geodesic Active Regions for Image Segmentation: a level set approach," *Proceedings of ECCV*, June 2000, Dublin, Ireland.

[62] C. Samson, L. Blanc-Feraud, G. Aubert, and J. Zerubia. "A Level Set Method for Image Classification," *Int. Conf. Scale-Space Theories in Computer Vision*, pp. 306–317, 1999.

[63] D. Snow, P. Viola and R. Zabih, "Exact voxel occupancy with graph cuts," *Proc. of the Intl. Conf. on Comp. Vis. and Patt. Recog.*, 2000

[64] S. Zhu and A. Yuille, "Region Competition: Unifying snakes, Region Growing, and Bayes/MDL for Multiband Image Segmentation," *IEEE Transactions on Pattern Analysis and Machine Intelligence*, **18**(9): 884–900, Sep. 1996.

LIST OF WORKSHOP PARTICIPANTS

- Mark Alber, Department of Mathematics, University of Notre Dame
- Madjid Allili, Department of Mathematics, Georgia Institute of Technology
- Yali Amit, Department of Statistics, The University of Chicago
- Ery Arias-Castro, Department of Statistics, Stanford University
- Amir Averbuch, Computer Science Department, Tel Aviv University
- Piero Barone, Istituto per le Applicazioni del Calcolo, Consiglio Nazionale delle Ricerche (CNR)
- Sukanta Basu, St. Anthony Falls Laboratory, University of Minnesota
- Sebastiano Battiato, AST - Catania Laboratory ST Microelectronics
- Peter Belhumeur, Electrical Engineering and Computer Science, Yale University
- Santiago Betelu, Institute for Mathematics and its Applications
- Irving Biederman, Neuroscience Program and Departments of Psychology and Computer Science, University of Southern California
- Mireille Boutin, School of Mathematics, University of Minnesota
- Thomas Carlson, Department of Psychology, University of Minnesota
- Michele Carriero, Department of Mathematics, University of Lecce
- Jamylle Carter, Institute for Mathematics and its Applications
- Vicent Caselles, Departamento de Tecnologia, Universitat Pompeu-Fabra
- Antonin Chambolle, CEREMADE, Universite de Paris-Dauphine
- Tony F.C. Chan, Department of Mathematics, UCLA
- Brian Chapeau, Department of Computer Science, Uninversity of Minnesota
- Rama Chellappa, Center for Automation Research, University of Maryland
- Li-Tien Cheng, Institute for Mathematics and its Applications
- Sing-Hang Cheung, Department of Psychology, University of Minnesota
- Alessandro Chiuso, Electronics and Informatics University of Padova
- Nicholas Coult, Department of Mathematics, Augsburg College
- Fabio Cuzzolin, UCLA

- Ingrid C. Daubechies, Department of Mathematics, Princeton University
- Boyue Dodov, University of Minnesota
- Ernst D. Dickmanns, Institut fr̈ Systemdynamik und Flugmechanik, Universität der Bundeswehr München
- David Donoho, Department of Statistics, Stanford University
- Grant Erdmann, Department of Mathematics, University of Minnesota
- Ian L. Dryden, School of Mathematical Sciences, University of Nottingham
- Fred Dulles, Institute for Mathematics and its Applications
- Selim Esedoglu, Institute for Mathematics and its Applications
- Olivier Faugeras, Sophia Antipolis Research Unit, INRIA
- Francois Fleuret, Projet Imedia, INRIA-Rocquencourt
- Efi Foufoula-Georgiou, St. Anthony Falls Laboratory, University of Minnesota
- Mirram Freedmen, University of Minnesota
- Davi Geiger, Computer Science and Neural Science Courant Institute, New York University
- Donald Geman, Department of Mathematics and Statistics, University of Massachusetts at Amherst
- Tryphon Georgiou, Department of Electrical and Computer Engineering, University of Minnesota
- Juan Gispert
- Jack Goldfeather, Mathematics and Computer Science, Carleton College
- Bernard Goulard, Centre de Recherches Mathematiques, Universite de Montreal
- Robert Gulliver, School of Mathematics, University of Minnesota
- Steven Haker, Department of Radiology, Surgical Planning Laboratory, Brigham and Women's Hospital
- Bruce Hartung, Electrical Engineering and Computer Science, University of Minnesota
- Thomas Heft, School of Mathematics, University of Minnesota
- John Hoffman, Maritime Surveillance Aircraft, Lockheed Martin
- Dirk Horstmann, Mathematisches Institüt, Universitat zu Koeln
- David Jacobs, NEC Research Institute
- Josef Keller, Department of Engineering, Tel Aviv University
- Daniel Kersten, Psychology Department, University of Minnesota
- Benjamin B. Kimia, Department of Engineering, Brown University
- Dimitri Kirill, Institute for Mathematics and its Applications
- Samuel Krempp, University of Massachussetts
- Christopher Lang, Indiana University Southeast

- Antonio Leaci, Department of Mathematics, University of Lecce
- Tai-Sing Lee, Center for the Neural Basis of Cognition and Department of Computer Science, Carnegie Mellon University
- Jean-Marc Lina, Centre de Recherches Mathematiques, University of Montreal
- Andrey Litvin, Electrical and Computer Engineering Boston University
- Bradley Love, Laboratory of Survival and Longevity, Max Planck Institute for Demographic Research
- Darek Madej, Advanced Development, Symbol Technologies
- Stephane Mallat, Ecole Polytechnique
- Bertalmio Marcelo, Electrical and Computer Engineering, University of Minnesota
- Riccardo March, Istituto per le Applicasioni del Calcolo (IAC), Consiglio Nazionale delle Ricerche (CNR)
- Rob Martin, University of Minnesota
- Andres Sole Martinez, Universitat Pompeu Fabra
- Massimo Mascaro, Department of Statistics, University of Chicago
- Donald E. McClure, Department of Applied Mathematics, Brown University
- Peter McCoy, Mathematics Department, U.S. Naval Academy
- Jason Miller, Division of Mathematics and Computer Science, Truman State University
- Willard Miller, Institute for Mathematics and its Applications
- David Mumford, Department of Applied Mathematics, Brown University
- Alexei Novikov, Applied and Computational Mathematics, California Institute of Technology
- John Oliensis, NEC Research Institute Inc.
- Peter Olver, School of Mathematics, University of Minnesota
- Stanley Osher, Department of Mathematics, UCLA
- Daniel N. Ostrov, Santa Clara University
- Javier Pascau, Medical Imaging Laboratory, Hospital Universitario Gregorio Maranon
- Eric Pauwels, Dept. PNA4, Centre for Mathematics and Information Science
- Victor Patrangenaru, Department of Mathematics and Statistics, Georgia State University
- Xavier Pennec, Unite de Recherche, INRIA Epidaure
- Pietro Perona, Department of Electrical Engineering, Caltech
- Ilya Pollak, School of Electrical and Computer Engineering, Purdue University

- Mary Pugh, Department of Mathematics, University of Pennsylvania
- Shantanu Rane, Electrical Engineering and Computer Science, University of Minnesota
- Anand Rangarajan, Department of CISE, University of Florida
- Christopher S. Raphael, Department of Mathematics and Statistics, University of Massachusetts, Amherst
- Kelly Rehm, Pet Imaging, VAMC
- Mohamed Ben Rhouma, Center for Dynamical Systems and Nonlinear Studies, Georgia Institute of Technology
- Fadil Santosa, IMA and MCIM
- Kirt Schaper, PET Imaging, Minneapolis VA Medical Center
- Otmar Scherzer, Institute for Applied Mathematics, University of Bayreuth
- Erik Schlicht, Department of Psychology, University of Minnesota
- Kevin Schweiker, Engineering Department, Freestyle Technologies, Inc.
- Jayant M. Shah, Department of Mathematics, Northeastern University
- Yonggang Shi, Electrical and Computer Engineering, Boston University
- Shuli Cohen Shwartz, U.C.G. Technologies Ltd., The Technion Enterpreneurial Incubator Co.
- Michael Israel Sigal, Department of Mathematics, University of Toronto
- Stefano Soatto, Department of Computer Science, University of California Los Angeles
- Nir Sochen, Department of Applied Mathematics, Tel-Aviv University
- Allen Tannenbaum, Departments of Electrical and Computer and Biomedical Engineering, Georgia Institute of Technology
- Bart M. ter Haar Romeny, Image Sciences Institute, University Medical Center
- Franco Tomarelli, Matematica Department, Politecnico di Milano
- Richard Tsai, Department of Mathematics, University of California, Los Angelesngeles
- Ben Tustison, St. Anthony Falls Laboratory, University of Minnesota
- Pradyumna S. Upadrashta, University of Minnesota
- Jean-Philippe Vert, DMA, Ecole Normale Superieure
- Curtis Vogel, Mathematical Science, Montana State University Hans Weinberger, School of Mathematics, University of Minnesota
- Alan S. Willsky, Department of Electrical Engineering and Computer Science, Massachusetts Institute of Technology

- Anthony Yezzi, School of Electrical Engineering and Computer Science, Georgia Institute of Technology
- Laurent Younes, Le Centre de Mathmatiques et de Leurs App., CNRS
- Alan Yuille, The Smith-Kettlewell Eye Research Institute
- Josh Zeevi, Department of Electrical Engineering, Technion-Israel Institute of Technology
- Haomin Zhou, Department of Applied Mathematics, Caltech
- Song Chun Zhu, Department of Computer and Information Sciences, Ohio State University
- Jiancheng Zhuang, Department of Radiology, University of Minnesota
- Steven Zucker, Computer Science and Electrical Engineering, Yale University

1999–2000 Reactive Flows and Transport Phenomena
2000–2001 Mathematics in Multimedia
2001–2002 Mathematics in the Geosciences
2002–2003 Optimization
2003–2004 Probability and Statistics in Complex Systems: Genomics, Networks, and Financial Engineering
2004–2005 Mathematics of Materials and Macromolecules: Multiple Scales, Disorder, and Singularities
2005-2006 Imaging

IMA SUMMER PROGRAMS

1987 Robotics
1988 Signal Processing
1989 Robust Statistics and Diagnostics
1990 Radar and Sonar (June 18–29)
New Directions in Time Series Analysis (July 2–27)
1991 Semiconductors
1992 Environmental Studies: Mathematical, Computational, and Statistical Analysis
1993 Modeling, Mesh Generation, and Adaptive Numerical Methods for Partial Differential Equations
1994 Molecular Biology
1995 Large Scale Optimizations with Applications to Inverse Problems, Optimal Control and Design, and Molecular and Structural Optimization
1996 Emerging Applications of Number Theory (July 15–26)
Theory of Random Sets (August 22–24)
1997 Statistics in the Health Sciences
1998 Coding and Cryptography (July 6–18)
Mathematical Modeling in Industry (July 22–31)
1999 Codes, Systems, and Graphical Models (August 2–13, 1999)
2000 Mathematical Modeling in Industry: A Workshop for Graduate Students (July 19–28)
2001 Geometric Methods in Inverse Problems and PDE Control (July 16–27)
2002 Special Functions in the Digital Age (July 22–August 2)
2003 Probability and Partial Differential Equations in Modern Applied Mathematics (July 21–August 1)
2004 n-Categories: Foundations and Applications (June 7–18)

IMA "HOT TOPICS" WORKSHOPS

- Challenges and Opportunities in Genomics: Production, Storage, Mining and Use, April 24–27, 1999

- Decision Making Under Uncertainty: Energy and Environmental Models, July 20–24, 1999
- Analysis and Modeling of Optical Devices, September 9–10, 1999
- Decision Making under Uncertainty: Assessment of the Reliability of Mathematical Models, September 16–17, 1999
- Scaling Phenomena in Communication Networks, October 22–24, 1999
- Text Mining, April 17–18, 2000
- Mathematical Challenges in Global Positioning Systems (GPS), August 16–18, 2000
- Modeling and Analysis of Noise in Integrated Circuits and Systems, August 29–30, 2000
- Mathematics of the Internet: E-Auction and Markets, December 3–5, 2000
- Analysis and Modeling of Industrial Jetting Processes, January 10–13, 2001
- Special Workshop: Mathematical Opportunities in Large-Scale Network Dynamics, August 6–7, 2001
- Wireless Networks, August 8–10 2001
- Numerical Relativity, June 24–29, 2002
- Operational Modeling and Biodefense: Problems, Techniques, and Opportunities, September 28, 2002
- Data-driven Control and Optimization, December 4–6, 2002
- Compatible Spatial Discretizations for PDE, November 3–7, 2003

SPRINGER LECTURE NOTES FROM THE IMA:

The Mathematics and Physics of Disordered Media
Editors: Barry Hughes and Barry Ninham
(Lecture Notes in Math., Volume 1035, 1983)

Orienting Polymers
Editor: J.L. Ericksen
(Lecture Notes in Math., Volume 1063, 1984)

New Perspectives in Thermodynamics
Editor: James Serrin
(Springer-Verlag, 1986)

Models of Economic Dynamics
Editor: Hugo Sonnenschein
(Lecture Notes in Econ., Volume 264, 1986)

Forthcoming Volumes:

Mathematical Systems Theory in Biology, Communications, Computation, and Finance

Transport in Transition Regimes

Dispersive Transport Equations and Multiscale Models

Geometric Methods in Inverse Problems and PDE Control

Mathematical Foundations of Speech and Language Processing

Time Series Analysis and Applications to Geophysical Systems